VOID

Foundational Questions in Science

At its deepest level, science becomes nearly indistinguishable from philosophy. The most fundamental scientific questions address the ultimate nature of the world. Foundational Questions in Science, jointly published by Templeton Press and Yale University Press, invites prominent scientists to ask these questions, describe our current best approaches to the answers, and tell us where such answers may lead: the new realities they point to and the further questions they compel us to ask. Intended for interested lay readers, students, and young scientists, these short volumes show how science approaches the mysteries of the world around us, and offer readers a chance to explore the implications at the profoundest and most exciting levels.

VOID

The Strange Physics of Nothing

James Owen Weatherall

Yale UNIVERSITY PRESS
NEW HAVEN AND LONDON

Templeton Press

Yale University Press books may be purchased in quantity for educational, business, or promotional use. For information, please e-mail sales.press@yale.edu (U.S. office) or sales@yaleup.co.uk (U.K. office).

Designed by Gopa and Ted2, Inc.
Set in Hoefler Text by Gopa and Ted2, Inc.

Printed in the United States of America.

Library of Congress Control Number: 2016949271
ISBN: 978-0-300-20998-3 (cloth: alk. paper)

A catalogue record for this book is available from the British Library.

This paper meets the requirements of ANSI/NISO Z39.48-1992 (Permanence of Paper).

10 9 8 7 6 5 4 3 2 1

To Cailin

Contents

VOID

Prologue

Much Ado about Nothing

IMAGINE A HOUSE with no furniture. Is it empty? Presumably I haven't given enough information to answer the question. There may be other stuff in the house: people, clothes, food, pets. Take all of this away, too. Indeed, take out all of the "stuff," big and small. Mop the floors, scrub the bathroom, and dust the window sills. Now is the house empty?

If this were a book about real estate, the answer would likely be "yes." At least, the house would be ready for a new owner to move in. But is it really *empty*? Perhaps there's a sense in which the rooms are empty, but of course the house still has some things inside. It has walls and floors, pipes and electrical wires, toilets and bathtubs and sinks. Take all of this out, too—gut the house completely, imagining, for the sake of argument, that we could do so without the house collapsing. We are left with just a shell of a house. But is it empty?

Of course not. There's still air inside. And that's not all. If the house is located in a reasonably well-populated town, it is probably near radio towers whose signals reach the house. Perhaps the neighbors have a wireless network with a similar effect. If the house has windows, then during the day it may be filled with sunlight. Radio waves and light rays are kinds of electromagnetic radiation, so there is radiation in the house. And since the house is on the surface of the earth, if one were to drop an apple, say, it would fall to the floor due to the earth's gravitational influence. In classical physics, at least, this would be because there is a *gravitational field* inside the house.

Fine, you think. This is getting a bit pedantic. But we can deal with it. Put the house deep in intergalactic space, far from stars or any other massive bodies. Let the air escape and shield the house from radiation of all sorts. Surely *now* the house is empty. There is nothing inside. Right?

There is a very old question, famous, or infamous, for its difficulty. *Why is there something rather than nothing?*[1] Part of what makes the question difficult is that it's not clear what could possibly count as a satisfactory answer. Explanations have to start somewhere; this question, however, seems to demand that we explain everything at once, without appeal to anything that *does* exist.[2] For just this sort of reason, many scientifically oriented philosophers—not to mention scientists—have dismissed the question entirely. It makes sense, these philosophers would say, to ask *what* there is, how it behaves, how we have come to be in the current state of the universe from earlier states. Not *why*.

We are accustomed to looking to physics for our answers to these latter sorts of questions, at least at the most fundamental level. And indeed, physics has yielded some impressive answers: we now know the material world is composed of such things as quarks and electrons, photons and gluons. The physics of *stuff* is well-trod territory. But what of the alternative? That is, this approach, of asking what there is and how it behaves, puts all of the emphasis on the *something* half of the question and ignores the *nothing* half. What, according to our best physical theories, is nothing? What would the world be like if there were no electrons, no quarks, no photons?

It is this last question that will be the focus of this book. My goal is to explain that if we want to understand *stuff*, if we want to use physics to study what there is in the world, we need to reckon with what the world would be like if there weren't anything at all. I call this the physics of nothing.[3]

The question might seem silly. Nothing, after all, would just be what we'd have if there weren't any something. In a sense, one might think, no matter what kinds of stuff exists or could exist, the situation if there were nothing would be the same: empty space, pure and simple. This was how the great seventeenth-century physicist Isaac Newton thought about things. He held that space could be thought of as an infinite container into which stuff could be placed or removed without affecting the structure of space itself.[4] It was a kind of theater in which physics would unfold. According to this picture, the physics of nothing is simple.

In fact, this idea may seem *so* obvious as to pass by without much notice. To say that if we removed all the stuff in the universe, we'd be left with empty space doesn't seem like a substantive physical assumption—much less one that could be false. But this is as wrong as anything in physics could be.

The twentieth century was a tumultuous time for physics. Centuries-old theories were overturned, first by Einstein's discovery of special and general relativity between 1905 and 1915; then by the development, during the mid-1920s, of quantum theory; and again during the 1940s, when these two new theories were combined into what is now known as quantum field theory.[5] Little was left unscathed by these revolutions—even the physics of nothing.

Indeed, understanding how the physics of nothing has changed with the advent of general relativity and quantum field theory is essential to understanding both what these theories tell us about the world and how dramatically they differ from classical theory.

This book traces these changes. The first chapter picks up at the beginning of modern physics, with the work of the physicist and mathematician Isaac Newton during the seventeenth century. As I explain, Newton's best-known contributions to physics concern his theory of the motion of physical objects, as summarized in the laws of motion that bear his name. His theory of motion still forms the bedrock for a modern physics education, three and a half centuries later. But what is perhaps less well appreciated is that in order to even formulate his laws of motion, Newton needed to redefine fundamental notions of space and time. This led to a radical new picture of what the world is like—and also, what it would be like if there were no stuff in the first place.

One of my emphases here is just how controversial Newton's ideas were at the time. I explain how Newton broke with earlier physicists, such as the Greek philosopher-scientist Aristotle, whose thought dominated European science for millennia, and René Descartes, a French physicist, philosopher, and mathematician who strongly influenced Continental science in the early part of the seventeenth century. Much of the chapter is devoted to a famous dispute about the structure of space (and, as it will turn out, time) between Newton and one of his most accomplished contemporaries: Gottfried Willhelm Leibniz, a German polymath and court librarian in Hanover. As we will see, what appears at first glance to be a debate about metaphysics and theology is really tracking very subtle issues about physics—specifically, what we need to assume about space and time for motion to be the sort of thing that could be governed by laws in the first place.

In the second chapter, I turn to the first of two revolutions in physics from the early twentieth century: Albert Einstein's theory of relativity. Einstein's theory may be understood as a revision of Newton's laws of motion, mandated by the success of James Clerk Maxwell's theory of light, electricity, and magnetism in the nineteenth century. Maxwell's theory was originally formulated in the context of Newtonian physics, but this soon led to conceptual problems and surprising failed experiments. Einstein realized that Maxwell had really taken the first step on a path to a much more radical reconfiguration of physics. Like Newton, Einstein understood that the new theory of motion he was describing required him to redefine basic notions of space and time, leading to a new picture of the structure of (empty) space.

Revising Newton's laws of motion and setting them in a new spatio-temporal framework better adapted to Maxwell's theory resulted in what we now call *special relativity*. But Newton had done something else: he had also provided us with a theory of gravitation. Newton's theory had lasted a long time, with a long litany of successes. It could explain (almost) all of the observed motions in the solar system; it could also explain basic terrestrial phenomena, such as the tides and the motion of cannonballs. But it appeared to be incompatible with Einstein's new theory of space and time. The search for a new, appropriately relativistic theory of gravitation took Einstein another decade. The end result was the theory that we now call *general relativity*.

General relativity is striking for many reasons. It is a theory in which massive objects affect the motion of other massive objects.

In this way, it is much like Newton's theory of gravitation. But whereas in Newton's theory this motion arises because bodies exert a gravitational force on one another, in general relativity bodies affect one another by changing the geometrical structure of space and time. This is a theory in which space and time are *curved*, in the same way that a beach ball or the hood of a sleek car might be curved. And just as rain will run differently down a curved surface, so too will planets move differently through curved space and time.

The idea that stuff—stars, books, blue whales—changes the geometry of space and time is a significant shift from Newton's theory, or anything that came before it. But Einstein's theory allows even more: it allows for space and time to be curved even when there is *nothing* present, anywhere or at any time. In fact, space and time can themselves behave in ways that are strikingly similar to electromagnetic radiation, such as light or radio waves. As we will see, this makes defining what it would even mean for there to be "nothing" in the universe extremely subtle.

In the third chapter, I look at quantum theory—particularly at quantum theory in the context of a relativistic understanding of space and time. Like relativity theory, quantum theory may be understood as a revision of Newton's laws of motion. But it is a *different* revision, and the program of bringing relativity theory and quantum theory together is very much incomplete. For instance, we still do not have an adequate quantum theory of gravity because of profound difficulties in making quantum theory compatible with general relativity. But as I discuss, the problems arise at a much earlier stage, and although our best attempt to formulate a version of quantum theory that is compatible with special relativity—a theory known as *quantum field theory*—has had some of the most remarkable empirical successes in the history of science, there remain important ways in which the theory is not well understood.

Even so, one thing that *is* well understood (we think) about quantum field theory is that the physics of nothing plays a special role. This is because possible configurations of stuff in the world are described in terms of how many particles there are—or rather, by how likely we are to detect a given number of particles in a certain kind of experiment. To talk about these possible configurations of stuff, physicists begin by defining a special state of the world—known as the *vacuum state*—that is meant to represent a situation in which there are *no* particles and then go on to say how that state would change if you added particles of various kinds.

So the idea of a quantum vacuum is fundamental. Which makes it particularly important that this state has some very surprising features, of a sort that further erode the once-clear distinction between "something" and "nothing."

In all of the theories I just discussed, the physics of nothing is remarkably rich. Exploring this richness offers insight into the foundations of these theories, in a way that emphasizes some of the most basic ways in which the worlds these theories describe diverge from one another—and in many cases, from our basic intuitions about nature. The physics of nothing boils these theories down to their bare essence, the core of their conceptual structure.

But as I explore in the epilogue, it reveals other things, too.

I suggested above that it is hard to imagine that a region of space with nothing in it could be different from what Newtonian physics tells us it should be. And yet for thousands of years before Newton, empty space was controversial, with many leading philosophers and proto-physicists declaring that the very idea was incoherent. Then, following the success of Newton's theory, the structure of space, time, and matter as understood in Newtonian physics began to be taken for granted—so much so that it became difficult to see how the world could be any other way. But it was precisely this "obvious" conception of empty space that was forced to yield in the face of new theories.

Of course, when our physical theories change, *something* about our conception of nature must change as well. And we should expect the basic principles of our best physics to evolve as we learn more about the world. But the point here is that before general relativity and quantum field theory, it was hard to recognize that there were substantial physical principles at work at all in our understanding of what it would mean to have a region of empty space. The physics of nothing seemed trivial—until we saw that it must be otherwise.

Indeed, according to general relativity and quantum field theory, empty space of the sort Newton imagined is physically *impossible*. Empty space is not merely a stage on which the physics of stuff can unfold; it has structure of its own that is every bit as interesting and complex as the structure of matter. Much of twentieth-century physics has been devoted to understanding just what that structure is.

The real moral of the story I tell in this book is that a crucial part of developing scientific theories is to take basic concepts—concepts like "nothingness"—and make them precise enough to support scientific inquiry. But this process of adapting our intuitive ideas to the more rigorous demands of science can result in radical changes to our conception of reality. And it is often these most basic concepts that are revised when we develop new theories.

The consequences of these sorts of changes are perhaps clearest in the context of the question I asked above. Why is there something rather than nothing? Does modern physics provide an answer to this question after all? In a sense, the answer seems to be "yes": as I just pointed out, modern physics—especially quantum field theory—tells us that nothing, in the sense we would usually have in mind, is impossible.[6] Empty space, at least on one conception, is ruled out by the laws of physics.

But something else is clear as well. How we understand this question and what might count as an answer to it depend on just what we mean by "something" and—perhaps even more—"nothing." You might have thought these terms were somehow unambiguous, independent of our physical theories or our other beliefs about the world. But the physics of nothing shows that this isn't right. How we understand nothingness can vary a great deal.

And this variation is important. How we understand nothingness in general relativity and quantum field theory goes right to the conception of reality that comes with these theories. Nothing really matters. This book tells you why.

CHAPTER 1

The Plenum and the Void

LATE IN the summer of 1665, as an epidemic of bubonic plague swept London, Cambridge University closed.[7] It would remain closed for two years. Scholars and students evacuated the city for the countryside to avoid contracting plague. A twenty-two-year-old Isaac Newton had just finished his bachelor of arts degree at Trinity College. Unable to continue his studies, he returned to his mother's home, Woolsthorpe Manor, in a small hamlet in the east of England.

The last time Newton had lived at Woolsthorpe Manor, he was seventeen years old. His mother had pulled him out of the King's School, the grammar school where Newton had been educated since he was twelve, to help her run the farm. A grammar school in the seventeenth century taught just that: grammar, or more specifically, *Latin* grammar. Newton had studied no mathematics, no science, and no philosophy—and now he was expected to learn farming. He was a slow study, and he hated the work.

Within a few months, the headmaster at the King's School had convinced Newton's mother that she had made a mistake, and young Isaac had better go back to his studies. At least, he was not going to make a living as a farmer. He returned to the King's

School, finished, and went on to Cambridge, supporting himself as a sizar—a kind of work-study position, where he served as valet for the wealthier, better-connected students. But he didn't mind: this was his chance to expand his intellectual horizons beyond Latin and farming to the best science, mathematics, and philosophy of the day. When Newton returned to Woolsthorpe during the plague year, he was a different man.

But he was poised to make an even more dramatic transformation. Left to his own devices in the countryside, he went from a student with undergraduate training to the leading mathematician in the world—though it would take some years before he was recognized as such. It was in Woolsthorpe that he single-handedly developed what he called his method of fluxions—the collection of mathematical methods we now call calculus, without which no modern mathematical science would be possible.[8] He also performed his first experiments with light and prisms, developing a theory of optics that greatly advanced that subject.

By themselves, these contributions would have made him one of the most important mathematicians and physicists in history. But it is for a third subject—one that would consume him for the next twenty years—that he is best remembered. While living in Woolsthorpe during the plague year, Newton began to think about motion and about gravitation.

Newton's laws of motion and his law of universal gravitation—which says that all bodies in the universe attract one another, with a force that increases with their masses and decreases with the square of their distances—revolutionized physics. Three and a half centuries later, we still teach Newtonian physics to high school and college students. Mastering Newton's work, and the work of his followers, is a prerequisite to any real understanding of physics—something that cannot be said for any prior physicist. Indeed, although we have subsequently developed theories that, in various ways, supersede Newtonian physics, Newton's laws remain a central pillar in our understanding of the physical world. For just one piquant example, : It was Newton's laws that took us to the moon, and that will take us to Mars.

At the heart of Newton's physics was a radical rethinking of our basic notions of space and time. Before Newton, our understanding of space and time was mired in metaphysical obscurity. Newton redefined these concepts in a way that allowed him to precisely describe various quantities of motion of physical objects, such as velocity and acceleration, and to relate these quantities to measurements we could make in a laboratory or observatory.[9]

Of course, Newton's goal in all of this was to develop the physics of *stuff*—ordinary objects like tables and chairs, as well as planets and stars. But before he could do that, he had to say what the world would be like if there were nothing. He needed to describe the geometrical structure of empty space.

Newton's masterpiece, the culmination of two decades of work on motion and gravitation, was the *Philosophiae Naturalis Principia Mathematica*—the *Mathematical Principles of Natural Philosophy*—or the *Principia* for short. It was first published in 1687, when Newton was forty-four years old. Although there are many documents, mostly unpublished during Newton's lifetime, that record the development of his thinking on motion—including a remarkable document known colloquially as *De Grav*, which many scholars take to show that Newton already had his mature understanding of space and time by the late 1660s—the *Principia* was the definitive statement of Newton's physics.[10]

In the physics of the *Principia*, there are *bodies* that, at any given instant, occupy *places* in *space*.[11] Bodies, for Newton, are extended objects—things like tables, chairs, and planets. Space, to a first approximation, may be thought of as an infinite, otherwise empty container in which bodies may be located.[12] Places are possible locations of bodies—literally, regions of space that could be filled by a body. We may think of space as all of the places—that is, all of the locations that could be occupied by a body at an instant. Between bodies, there is nothing but empty space.

Newton's picture of bodies in space traces back to the ancient Greeks, most notably Democritus and Epicurus, who developed variants of a theory they called *atomism*.[13] Atomism holds that the basic constituents of matter are tiny, indivisible bits of matter known as "atoms."

Our contemporary notion of atom, as it appears in twenty-first-century chemistry and physics, owes much to this ancient idea, but it differs in important ways. The biggest difference is that, for the Ancients, atoms were indivisible. They did not have parts. Atoms as we understand them now *do* have parts. They are composed of electrons, protons, and neutrons; protons and neutrons, meanwhile, are composed of still more basic entities known as quarks. And atoms are certainly divisible: dividing them is the basis of nuclear fission, as used in nuclear weapons and nuclear power.[14] Another difference is that the Ancients believed atoms could differ in properties such as size, shape, and color—though there was apparently no attempt to systematically classify atoms. Twenty-first-century chemistry, meanwhile, recognizes a periodic table of different atoms, with some 118 varieties, but the differences in species of atoms is determined by how they are composed. It would be very strange to think of modern atoms as having colors or shapes in the way the Ancients imagined—though they do, arguably, differ in size.

Newton believed that bodies have smallest parts, which could not be further divided, so he believed in something like the Ancients' atoms.[15] But for our purposes, atoms are not the important part of the atomistic view. Rather, it is that for Ancient atomists, atoms move around in the *void*, infinite space occupied by nothing. All physical phenomena, they believed, could be explained by motions of atoms in the void: tiny bits of stuff moving around in regions without stuff.

This idea that physical stuff comes in discrete chunks (indivisible or not), and that between bits of matter there are regions of space where there is nothing, may seem intuitively plausible or even obvious. After all, we are used to thinking about distinct things having some space between them. If you imagine sitting at a café with a friend, it is very natural to think that there's you, your friend, and in between, some distance of space where there's nothing. Of course, that is not literally right, since there is air between you and your friend. But we are taught that air is composed of various molecules, such as carbon dioxide and diatomic oxygen. And these molecules have nothing in between them.

If this idea does seem intuitive, it is because we are brought up in a world shaped by three and a half centuries of Newtonian science. But for millennia before that, the idea was extremely controversial.[16] Aristotle, for instance—by far the most influential physicist before Newton—argued that Democritus' views were not only wrong but incoherent. Motion, Aristotle claimed, could only arise from pushing or pulling of one thing by another. Motions that are not obviously of this form—smoke rising, rocks falling—occur because there is a natural order between different kinds of matter, and when matter is arranged out of that order, neighboring bits push and pull one another to restore the natural order. So a rock suspended in air and released is pushed downward by the air to join the other rocks on the ground, while the rock pushes up the air around it.

The important thing about this picture is that it precludes motion in the void. If an atom were surrounded by nothing at all, there would be nothing to exert any sort of pushing or pulling influence on it. If all motion arises from pushing and pulling by neighboring matter, and there is no neighboring matter, then motion is impossible. The upshot was supposed to be that the very idea of empty space is confused.

Aristotle's physics dominated the science of the Western world for nearly two thousand years.[17] By the time Newton was born, however, Aristotle's influence was already on the wane. Aristotelianism had been challenged in the late sixteenth and early seventeenth centuries, by members of the vanguard of the scientific revolution, such as Galileo Galilei in Italy and René Descartes in France. But the new views that emerged in the wake of Aristotelian physics—at least in Continental Europe—were no more sympathetic to the idea of bodies moving through empty space than Aristotle had been.

Descartes was one of the most influential natural philosophers working in the first half of the seventeenth century.[18] He rejected Aristotelian physics and attempted to replace it with a complete, systematic physics of his own. Part of this project involved developing a complete metaphysics, with accounts of space and time, God, and matter of all sorts. He was an early advocate of the idea that an adequate philosophy of nature must be based on mathematics. This approach led him to develop a theory of space, time, and matter based on principles of geometry. For Descartes, the primary property that an object had was its *extension*, that is, its shape in space. Space, meanwhile, was nothing but extension—the shapes of the bodies occupying space.

This tight connection between stuff and the space filled by stuff—so tight, in fact, that there was no difference between them as far as Descartes was concerned—led to a very quick argument that empty space is impossible.[19] Suppose there were a void—some region of space that was empty of all matter. Then this region would have to have some extension: it would have to have nonzero volume, or else it would not be properly described as a region in the first place. But since extension simply *is* body, there would have to be some sort of body present in the region—that is, it would not be empty after all, by definition of being extended!

Thus, Descartes reasoned, all of space must be filled with some sort of very fine, unobservable stuff. He called this material the *plenum*. Ordinary objects can pass through the plenum, but anywhere that is not occupied by stuff like tables, chairs, and planets is occupied by this other sort of stuff instead.

If you think this idea seems strange, or even question-begging—after all, Descartes *defines* space in such a way that it can never be empty—then you are in good company: Descartes' views on space and matter were flatly rejected by the leading physicists of the second half of the seventeenth century.[20] But as we will see, versions of this idea have resurfaced over the subsequent centuries. Arguably, even today, our best current physical theories posit a plenum—and at the very least, developments in twentieth-century physics have significantly blurred the distinction between "space" and "stuff" occupying that space, albeit in different ways from what Descartes envisioned.

In a letter dated February 5, 1676, Newton penned one of his most-quoted remarks: "If I have seen further, it is by standing on the shoulders of Giants."[21] He cannot take credit for this metaphor, which already had a five-hundred-year history by the time he wrote his letter—usually with the added twist that it is a dwarf benefiting from the giants' height. And though the quote is often cited in connection with scientific humility, the passage may not be quite as modest as it seems: the letter was addressed to Robert Hooke, who was short and humpbacked, and the "Giants" to whom Newton referred were Hooke and Descartes, for whom he had little admiration. Indeed, by the time of writing, Newton and Hooke were serious rivals, disagreeing bitterly over the nature of light. Newton had even threatened to resign from the Royal Society of London, the most important scientific organization in England at the time, over Hooke's criticisms of his work. Years later, Hooke, who was famously cantankerous, would claim that Newton had stolen his own theory of universal gravitation. With this context, it is hard not to see Newton's reference to giants as tongue in cheek.

Still, Hooke's stature notwithstanding, it is a remarkable fact that among Newton's contemporaries were a number of physicists, mathematicians, and natural philosophers whose abilities and contributions rivaled his own. One of the most influential was Gottfried Wilhelm Leibniz.[22] Two years younger than Newton, he was born in 1646 in the city of Leipzig, in the east of what is now Germany but was then part of the Holy Roman Empire. He spent most of his career as court librarian in Hanover, the seat of the Duchy of Brunswick-Lüneberg.

Though he would go on to develop systematic theories of physics, and to invent calculus at about the same time as Newton, Leibniz was not formally trained in mathematics or natural science. His only formal degrees were in law—training he put to considerable use as a political advisor to the Dukes, and later Electors, of Brunswick-Lüneberg. (In fact, Leibniz's research was instrumental in elevating the Duchy to an Electorate in 1692.[23]) Although he attended a leading preparatory school in Leipzig and later went on to study at the Universities of Leipzig (at fourteen years old) and Altdorf, in Nuremberg, Leibniz always described himself as an autodidact.[24]

As he would later write in autobiographical notes, much of his training came from hours spent locked away alone as a small child in his father's massive library, which was opened to him at eight years old at the urging of a local nobleman, over the strenuous objections of Leibniz's schoolteachers. (Leibniz's father, a professor of moral philosophy at the University of Leipzig, died in 1652, when Leibniz was six.)

Leibniz had a voracious intellect and a remarkably open mind. He studied everything, from politics and history to theology, biology, geology, mathematics, physics, and, of course, philosophy. He was willing to entertain any idea—to study it, criticize it, and most importantly, try to take something of value from it.[25] For instance, in the late 1660s, around when he finished his doctorate—and while Newton was holed up in Woolsthorpe—Leibniz became intrigued by alchemy. In the seventeenth century, alchemy was still perceived as a respectable area of study in many quarters, as much proto-chemistry as mysticism. (Newton, too, spent many years studying alchemy. Indeed, he took it far more seriously, for much longer, than Leibniz ever did.) And the claims of the initiated were truly remarkable: alchemists wrote that they had discovered profound hidden truths about the nature of the world and that the adept could perform various sorts of magic.

These claims were met with skepticism among the more hard-headed Enlightenment scholars, Leibniz included. But he also felt that if what they said was true, it was worth learning, and in any case, he should not cast judgment before getting better acquainted with the subject.

So Leibniz pored over alchemical lore, studying everything he could find. But as he read more, the subject only became murkier. He gradually came to a startling hypothesis: *none* of the people writing alchemical tracts had any idea what they were talking about. To test this, he wrote a letter to a local society of alchemists, mimicking what he took to be the most obscure passages from the texts he had read. As far as Leibniz could tell, the document he produced was completely unintelligible. But it was received by the local alchemists as a work of great learning—evidence that Leibniz was a master. He was invited to join the society and even offered a stipend to support his alchemical research. This seemed like firm evidence that alchemy was a sham, and Leibniz soon lost interest.

Leibniz brought a similar mix of open-mindedness and healthy skepticism to his many other pursuits, including physics. By the time he was writing, Cartesianism—the tradition of Descartes and his followers—was ascendant in both Continental science and philosophy. Leibniz was fiercely critical of Descartes, particularly on the metaphysics of physical objects. And while Descartes himself and most of his contemporaries had been critical of Aristotle and the tradition of Aristotelianism in Western Europe, Leibniz often tried to reconcile Aristotelianism with the features of Cartesian philosophy that he found most attractive.[26]

He had a penchant for finding compromises between traditionally warring intellectual traditions. His views on the possibility of empty space are a characteristic example.[27] Unlike Descartes or Aristotle, Leibniz believed empty space was possible, in the sense that one could perfectly well imagine a region of space where there isn't any stuff. In this, his views were similar to Newton's.

But unlike Newton, Leibniz held that the *actual* world could not have any such empty regions. Leibniz believed God could only have created the *best* of possible worlds (for otherwise, God must have decided to create a world that was not as good as possible, which Leibniz thought would be inconsistent with God's infinite goodness). He reasoned that if there were a region of empty space in the world, that region could have been filled—and by filling it with something intrinsically better than nothing, God could thus have produced a better world. So Leibniz concluded that all space must be filled, which led Leibniz to believe, much like Descartes, that there must be a plenum. In this way, Leibniz's views looked much more like Descartes' than Newton's.

By the time the *Principia* was published, Newton had been Lucasian Professor of Mathematics at Trinity College, Cambridge, for nearly twenty years. This was—and is—a distinguished position, befitting his status as the leading mathematician and physicist in the world. (More recent Lucasian professors have included Stephen Hawking and Michael Green, one of the inventors of string theory; Newton was elected to this position in 1669, only one year after receiving his MA degree. He was twenty-six years old.) Following the publication of the *Principia*, Newton catapulted to international stardom.[28] Leibniz quickly took notice.

The most serious dispute between Newton and Leibniz during their lifetimes had nothing to do with space—or even physics. It concerned calculus, the field of mathematics that Newton developed during the plague year in Woolsthorpe—and without which the *Principia* would not have been possible.[29] The trouble was that although Newton first developed calculus in the mid-1660s and circulated a small handful of documents around that time to mathematicians in Cambridge, he did not publish anything on the topic until much later. Even the *Principia*, although it implicitly used methods from calculus, did not include a detailed or systematic treatment. It was only in 1693 that Newton published a full account.

In the meantime, Leibniz had published several detailed articles on calculus, beginning in 1677, using a radically different notation. His notebooks show that he had begun thinking about the subject in 1675. Of course, this was already a decade after Newton had developed very similar ideas, and it seems no one ever disputed whether Newton got to calculus first, nor whether he did so independently of Leibniz. The quarrel was over whether Leibniz discovered calculus *independently* of Newton, or whether, as Newton and his followers would later assert, Leibniz had somehow gained access to one of Newton's early, unpublished (but circulated) manuscripts, and stolen the ideas.

The strange thing about this dispute is that it only took place in the early eighteenth century, decades after Leibniz's papers appeared. When Leibniz first published on calculus, in the 1670s, no one in England objected. Even in the 1690s, when Newton did finally publish his work, he and Leibniz both seemed to acknowledge one another's contributions, and their contemporaries, such as the French mathematician Guillaume François Antoine de l'Hôpital, who published a book on calculus (in Leibniz's style) in 1696, attributed the subject to both Newton and Leibniz.[30]

It was not until 1705, when an anonymous review of an appendix to Newton's *Opticks* appeared, that things began to get ugly.[31] The appendix was an extended treatment of topics from calculus that were relevant to the physics contained in the book. The reviewer, who turned out to be none other than Leibniz himself, spent several pages comparing Newton's contributions to Leibniz's. His tone seems mild enough today, but at the time, he was widely interpreted as implying that not only was Newton's work less significant than Leibniz's own, but also that Newton had spent years improving his understanding by studying Leibniz's published writings and that the material Newton ultimately published in the 1690s and 1700s was much more strongly indebted to Leibniz than Newton acknowledged.

By the time Leibniz's review appeared, Newton was living in London. In the early 1690s, he had suffered a nervous breakdown—possibly as a result of mercury poisoning, due to his experiments in alchemy.[32] Then, in 1696, he was appointed warden of the Royal Mint in London, and he left Cambridge, convinced it had become an intellectual backwater. (He did not formally resign his professorship until 1701.) He was elected president of the Royal Society in 1703, and he held both positions until his death in 1727. Newton continued to do some scientific work while in London, including publishing several major revisions of the *Principia*, but by this time the work for which he is best known had already been done, and much of his attention was devoted to other matters, such as a major recoinage and various theological projects.[33] Even so, his positions in London were even higher profile than the position in Cambridge, and Newton's political and scientific influence only grew.

Thus, Leibniz's review attacked not only an important English scientist but also a major political figure and, as president of the Royal Society, the de facto leader of the English scientific establishment. It was a major affront. One of Newton's followers, a Scottish mathematician named John Keill, soon wrote a response, alleging, in turn, that Leibniz had plagiarized Newton. This led to a flurry of letters and replies, until finally, in 1711, Leibniz formally asked the Royal Society of London to intervene and settle the matter.

Of course, Newton was then president of the Royal Society. And the scholars assigned to investigate were avowed Newtonians, already embittered by Leibniz's accusations. The document they produced in 1712 carefully detailed the timeline of Newton's contributions, finding that he had fully and independently developed the whole of calculus, with no significant contributions from Leibniz. In the wake of this document, communication between Continental and English mathematicians broke down. Relations remained seriously strained for much of the eighteenth-century.

It was with this bitter dispute in the background that Leibniz wrote, in 1715, to Caroline of Ansbach.[34] By that time, Caroline was the Princess of Wales—though Leibniz had known her long before that. In 1701, Parliament had passed the Act of Settlement, which stated that a Catholic could not ascend to the English throne. This law effectively barred any of the surviving members of the Stuart dynasty, which had ruled England for a century and Scotland for over three centuries, from producing another king or queen of a newly united Great Britain.

Thus, when the last Protestant Stuart, Queen Anne, died in 1714, the crown passed to her nearest Protestant relation, who happened to be her second cousin, George of Hanover, the Elector of Brunswick-Lüneberg—a position Leibniz had helped him achieve. When George became George I of England, Leibniz maintained his relationship not only with the king but also with his son, George Augustus, and with Caroline, George Augustus' wife, whom Leibniz had tutored since 1704.

It was thus with some interest that Caroline read, in Leibniz's letter, of the pernicious influence of Newton's work on theology in England. Alarmed, she passed the letter along to an English philosopher and theologian, Samuel Clarke. Clarke, who had been a chaplain in the royal household under Queen Anne, was known at court; moreover, as rector at Newton's parish church, St. James' in Piccadilly, he had recently come to know Newton personally—though he had been an outspoken defender of Newtonian ideas for decades before that. Clarke replied to Leibniz, initiating a correspondence that lasted until Leibniz's death in November 1716.[35]

Although Clarke was the nominal correspondent, and later published the letters under his own name when Leibniz died, it is certain that Newton had read Clarke's drafts and provided guidance before they were sent.[36] (Some historians have even speculated that Newton was the true author.) The correspondence touched on many topics, mostly theological and metaphysical. But the most influential passages connect only tangentially to the dominant themes. It was in his letters to Clarke that Leibniz mounted his most important critique of Newton's views on space and time—and, with Newton's guidance, Clarke offered replies.

To best understand Leibniz's criticisms of Newton's picture of space, it is best to begin with what they agreed on. Newton and Leibniz agreed completely on the *geometry* of space—that is, they agreed on what sorts of geometrical relations there are between places at any given instant of time. This geometry can be characterized, somewhat anachronistically, in the following way. First, for both Newton and Leibniz, space consists of a collection of possible locations of bodies—what Newton called places. We will call infinitely small places, that is, possible locations of infinitely small bodies, *points* of space. Space, then, consists of an infinite collection of points.[37]

These points are related to one another in various ways. For instance, given any two points, there exists a straight line that runs from one point to the other. We can think of this straight line as an arrow, beginning at one of the points and ending at the other. This straight line can be assigned a number, which we think of as a *length*, that represents the *distance* between the two points connected by the line. More generally, we can imagine drawing curved lines between any two points and using the distances between all of the points along these curved lines to calculate a length for the curved line. If we do this, it will turn out that a straight line between two points is the *shortest* line between the points. There is also a relation that can hold between two straight lines starting at the same point: they have an *angle* between them.

Now suppose I pick a point of space. (It doesn't matter which one: what I am about to say holds for any point.) Call this point p. Now imagine an arrow, or straight line, that begins at p. Call it v. As I have set things up so far, this arrow has to end somewhere—say, the point q. There are a few things I can do with the arrow v. One thing I can do is *stretch* it or *shrink* it. This means I can make the length of the arrow longer or shorter, or even turn it around so it points in the opposite direction. If I do this, the arrow I end up with will still start at p, but it won't end at q any longer; instead, it will end at some other point, q′. This gives us a way of capturing the idea that space is *infinite*.[38] Given any arrow v, at any point p, I can always stretch v as far as I like, and there will still be a point q′ for that arrow to point to. In other words, there are points that are as far away from p as I like.

A second thing I can do with the arrow v is *add* to it another arrow that begins at p. Imagine a second arrow, u, that also begins at p. Now imagine moving that arrow, without changing its length or direction, to q, the point where v ends. This new arrow, u′, which begins at q, must end at some point, which I will call r. And since p and r are just two points in space, there is an arrow that begins at p and ends at r. This arrow is what I get by adding v and u.

These operations of stretching, shrinking, and adding arrows let us define something else that Leibniz and Newton agreed on: they agreed that space is *three-dimensional*. What this means is that, at any point of space, I can always find three arrows, call them x, y, and z, with the following two properties: (1) there is no way to construct the arrow x (or, respectively, y or z) by any process of shrinking, stretching, or adding the other two arrows, y and z (and, respectively, the other pairs); and (2) *any* arrow that begins at p *can* be constructed by shrinking, stretching, and/or adding all *three* of these arrows. We can think of x, y, and z as pointing in three different, independent directions, in a space where there are no *other* independent directions for a fourth arrow to point in. Space is three-dimensional because there are three such directions—no more, and no less.

So Newton and Leibniz agreed on this much: Space is infinite and three-dimensional, any two points can be connected by a straight line, and there are objective distances between points and angles between straight lines. But if they agreed on all this, what could the Leibniz-Clarke dispute have been about?

Leibniz's most famous argument in the Leibniz-Clarke correspondence was directed at a crucial set of distinctions that Newton had drawn in the *Principia*, when he defined what he meant by space and place. Recall that for Newton, at any instant, any given body occupies some place in space, where space consists of all of the places a body may occupy. When he introduced these terms, Newton distinguished between *absolute* space and place on the one hand, and *relative* space and place, on the other.[39] When we use words like "here" or "there," we are referring to relative places—relative, that is, to a particular speaker. When we measure distance, we are measuring these quantities relative to some particular standard: say, a ruler. We do not have ways of describing locations or distances without using some sort of reference.

Still, according to Newton, we can at least conceive of space and place without referring to any particular observer or measuring apparatus—that is, from a kind of God's-eye view. This is what he means by "absolute" space and place. Our determinations of relative place are then to be understood as (imperfect) descriptions of these absolute notions. So when I say that a given chair is about two feet away from here, or that the sun is about 93 million miles away from earth, I am using relative places to say something about absolute places. Absolute space consists in all of the *absolute* places, that is, all of the locations of things, not from any given perspective, but as they are independent of any observer or measuring apparatus.

Leibniz thought that the very idea of absolute space was metaphysically incoherent.[40] (He also rejected absolute time, which Newton defended.) To support this, he offered what is now called the *shift* argument.[41] Suppose, for the sake of argument, that you believed in absolute space. Imagine the whole world at an instant. At that instant, everything—every body—must occupy some place, that is, some region of absolute space. Now suppose that, instead, all bodies in the universe were *shifted* by, say, five feet in some direction—say, along the line connecting the center of the earth to the center of the sun. (Of course, the earth and sun are also shifted.)

Since *everything* has been shifted, all of the relative places would stay the same: that is, none of the angles or distances—the geometrical quantities that both Newton and Leibniz agreed made sense, and could be measured—would have changed. And yet, since everything has been shifted, every single object would occupy a different *absolute* place. In other words, you could change the location of everything in the universe relative to absolute space without making any changes that could be measured or even described using any of our ordinary methods. Thus there would be two distinct possible configurations of all the stuff in the universe—actually infinitely many, corresponding to all of the possible shifts—according to which there could be no measurable or otherwise physically relevant difference.

According to Leibniz, this argument revealed two profound problems with Newton's idea of absolute space. First, Leibniz believed that if two physical situations are indistinguishable from one another, even in principle, then they must be the same. Roughly, if two things are to be different, there must be some *way* in which they are different; that is, they must differ in some property. He called this idea the Principle of the Identity of Indiscernibles. But in this case, Leibniz claimed, we have two situations that are not distinguishable. Thus, he concluded, they cannot actually be distinct situations at all—and so, absolute space must not be real.

The second problem ran even deeper. If, as Leibniz argued, the existence of absolute space implied that the original universe and the shifted universe did not differ in any physically significant way, then God must have made a choice in creating the universe, concerning *where* in absolute space to put all of the bodies in the universe. (This is how Leibniz's arguments about space relate to the theological issues taken up in his correspondence with Clarke.) And this choice would have been completely arbitrary: God could have chosen differently, without making any difference in any physical quantity. But this was impossible for Leibniz, because it violated a second of his principles, which he called the Principle of Sufficient Reason. This principle stated (again, roughly) that God could not do anything without a reason for doing so: to act arbitrarily would be inconsistent with God's infinite rationality. And so again, absolute space was impossible.

It is easy to imagine, when considering Leibniz's argument in isolation, that the disagreement came down to a combination of theology and metaphysics, which one would expect to matter little for physics. But carefully considering Clarke's reply to Leibniz shows where the real heart of the disagreement lay. It also shows why Newton introduced the notions of absolute space and absolute place in the first place: far from dabbling in metaphysics, Newton believed he needed these notions for the physics of the *Principia* to make any sense.[42]

Clarke does not offer anything truly new in response to Leibniz. His most convincing rejoinder was, essentially, to point to a crucial passage in the *Principia*.[43] This passage comes at the end of the section where Newton introduced the relative/absolute distinction, but it does not directly concern absolute space or place. Instead, it concerns absolute *motion*.

As I said above, Newton's real goal in the *Principia* was to present laws governing how bodies move and to explore the consequences of those laws. He gave three laws of motion, which form the foundation of Newtonian physics. These may be stated as follows: (1) a body will move in a straight line, at a constant speed, unless it is acted on by an external force; (2) the force exerted on a body is equal to its mass times its acceleration (i.e., F=ma), where acceleration measures how much a body is deviating from straight-line, constant speed motion; and (3) for every force exerted on one body by another, the second body exerts an equal and opposite force on the first body. Newton also introduces the law of universal gravitation, which accounts for some of the motion we see in the world. Gravitation is just one of the forces that are governed by Newton's laws of motion.

Here is the crucial point, though. In order to even write down these laws, Newton needed to define what he meant by motion. Motion, after all, along with force, is a basic concept that appears throughout the laws.[44] And to do this, he introduced another absolute/relative distinction: that between absolute motion and relative motion. Relative motion, like relative place, is motion as determined by some particular reference body: say, an observer or a system of tools for making measurements.

Suppose you are driving in a car on the highway, and another car is passing you on the left. Your car is moving relative to both the road and the car that is passing you. You, meanwhile, are *not* moving relative to your car—but you are moving relative to the road and the other car. All of these are relative motions.

Absolute motion is motion that is independent of any particular reference body—or, if you like, it is motion relative to absolute space. As far as Newton's physics is concerned, *all* of the things described in the last paragraph are undergoing absolute motion, even though some, such as you and your car, are not moving relative to one another.[45] This is because the road, which is at rest relative to the earth, and all of the cars driving on it are spinning around the earth's axis, while the earth is orbiting around the sun, the sun is orbiting around the center of the Milky Way, and so on.

It is *absolute* motion that is governed by Newton's laws of motion. To see this, just consider Newton's first law. Suppose we applied this to the case of the cars just described. (If you are worried about friction, then imagine they are rockets, deep in intergalactic space.) The relative version of the law would say that the car passing you on the left would have to continue on in a straight line at a constant speed, relative to your car, unless acted on by some outside force. But that's simply not true: there is an easy way to change the relative speed of your car and the person trying to pass you: you can slam your feet on your own brakes! In this case, you apply a force to your own car—but the relative speed of the other car changes, too. Indeed, it is hopeless to come up with laws of motion that govern arbitrary relative motion, because the state of motion of the reference body could change arbitrarily.[46] To make the law work, it cannot be about relative motion in general. It needs to be about a special kind of motion, which is what absolute motion is supposed to provide.

So Newton needed to define absolute motion in order for the laws of motion to make sense. But to define absolute motion, he believed, he needed to define absolute place, because absolute motion, for Newton, was *change* in absolute place over (absolute) time. This is what it meant to have motions relative to absolute space.

How much of this Leibniz understood is not clear. He remained skeptical of various parts of Newton's physics, but in his own attempts to develop a physical theory, he seems to have accepted a kind of absolute motion—and also, like Newton, he took straight-line, constant-speed motion as the "default." But he never gave a coherent story about how to make sense of this sort of motion without absolute space.[47] So it seems he wanted it both ways: on the one hand, he wanted a concept of absolute motion suitable for developing laws of motion, and on the other, he wanted to reject any notion of absolute space.

Of course, Leibniz's arguments against absolute space could

be convincing even if Leibniz never provided a compelling alternative definition of motion that could do without it. If so, then at the very least there was *something* wrong with Newton's idea of absolute motion. In fact, more is true: If absolute motion in Newton's sense is undetectable, like absolute place, then Leibniz could make the same arguments against absolute motion that he gives against absolute space. For instance, instead of shifting the whole world by five feet to the left, suppose you set everything in the universe in motion, so that all of the *relative* motions were the same, but now everything was moving at some fixed speed in a fixed direction, relative to absolute space.[48]

This possibility was one that Newton anticipated in the *Principia*—and it was to these passages that Clarke directed Leibniz. Newton had an argument that at least *some* absolute motion *was* detectable. His example has become very famous in the history of physics.[49] Suppose you have a bucket filled with water. If the bucket is sitting on the earth, stationary relative to the earth's surface, the surface of the water would be (more or less) flat. But now suppose you begin to rotate the bucket around the axis running top to bottom through the center. As the bucket starts moving, the surface of the water will change shape: the edges nearest the outside will rise, so that the whole surface is shaped like a shallow bowl.

In this case, of course, we are talking about a bucket that is rotating relative to the surface of the earth. But Newton went a step further. Suppose the universe was completely empty, except for a single bucket filled with water. Then, he claimed, there would still be a difference between the case where the bucket was at rest and the case where it was moving; one would be able to tell just from the shape of the surface of the water in the bucket. But now there is nothing at all in the universe relative to which the bucket could be moving. What could explain the difference, then, between the different shapes taken by the water's surface? According to Newton, the difference came down to different states of absolute motion. And because he defined absolute motion by way of absolute space, he claimed, absolute space *did* have physical significance after all.

In a landmark paper published in 1967, three hundred years after Newton's remarkable plague year, the twentieth-century philosopher Howard Stein made a startling observation that significantly changed how philosophers and historians of science understand the *Principia*.[50] What Stein realized is that if the core of the disagreement between Newton and Leibniz really concerned what sort of structure one needs to develop a successful physical theory of motion, then their dispute was not really about the structure of *space* at all; rather, it was about the *relationship between space and time*. It was a dispute about whether it makes sense to talk of the *same place* at *different times*.

Of course, there is one obvious sense in which we often do talk about the same place at different times—for instance, I can talk about my house, or the local grocery store, or the Commonwealth of Virginia at various times. If I say I have lived in this house for several years, or I go to the grocery store every week, or that today Virginia is a state in the United States, but in the early eighteenth century it was a British colony, there is no ambiguity about what I mean.

But as we have seen before, there is another way of looking at these matters. After all, as with the road in the car example above, all of the places just mentioned are on the surface of the earth, and we believe the earth is moving. So although there is one sense—

really, a relative sense, since what I care about is the location of my house relative to various other landmarks, and the grocery store relative to my house, and Virginia relative to North Carolina and Maryland—in which these are the *same* place at different times, there is also a second sense in which "my house," "the grocery store," and "Virginia" do *not* refer to the same places at different times.

The basic question, then, is whether this second sense is really distinct from the first—or if it even makes sense to say that "really" Virginia has moved. In other words, is the situation merely that while my house does not move relative to the houses around it—and in this sense, stays in the same place over time—it does move relative to some reference bodies, such as the sun, Mars, and the center of the Milky Way, and in this sense does *not* stay in the same place over time? Or is it that, while all of these relative senses of change of place are perfectly fine, there is nonetheless some basic fact of the matter about whether something is in the same place at two different times—irrespective of whatever relative changes in place have occurred? As you can perhaps guess, Newton thought there *was* such a fact of the matter, even though we could never tell whether any particular body in fact occupied the same place at different times, and Leibniz thought that since relative motions were all we could ever determine, there was *no* fact about whether a given body occupied the same place, or some different place, at different times.[51]

So far, this is just window-dressing: thinking about sameness of place over time simply recasts the debate described above in somewhat different terms. But this new way of thinking leads to a deeper insight. As I noted above, Newton and Leibniz agreed on the geometry of space, but they did *not* agree on the geometry of *space-time*.

What do I mean by "space-time"? As I explained above, *space* may be thought of as a collection of points related by straight lines, or arrows. *Space-time* may be thought of in the same way, as points related by straight lines, except that whereas space is three-dimensional, space-time is *four*-dimensional. This means that at any point p of space-time, I can find *four* arrows—call them x, y, z, and t—with the properties that (1) none of these four can be constructed by any process of shrinking, stretching, or adding the other three, and (2) I can construct any other arrow by stretching, shrinking, and/or adding together x, y, z, and t. We can think of this four-dimensional structure as arising from infinitely many copies of space stacked one on top of another in a fourth dimension, much as a stack of (approximately) two-dimensional sheets of paper forms a three-dimensional book.

Recall that points in ordinary space were supposed to be locations that very small bodies might occupy—infinitely small places. But what is a point in space-time? It is a location in space *and* time. The things that occupy spatio-temporal locations of this sort are sometimes called "events." An *event* is something that occurs in a small region of space, at an instant of time. Snapping your fingers, say. Or a first kiss.[52] Note that neither objects that are extended in space (say, a rope) nor objects that persist through time (say, a particle) are represented by single points in space-time. Instead, these are represented by various sorts of curves and surfaces, composed of many different events.

We saw above, when discussing space, that points are related by distances. Analogous relationships hold between points in space-time. One such relationship is *simultaneity*: two events may occur at the same time, as when I snap two pairs of fingers at once. More generally, nonsimultaneous events occur in some time order: first one happens, and then later another does. Finally, I have a notion of *duration*—that is, distance in time—between events at different times, so I can say how much time has passed between two events. I can think of any maximal collection of events that are mutually simultaneous as corresponding to a copy of space at a given time, and then the temporal ordering of events lets me think of these copies of space as arranged one after another, at successive times.

Each of these copies of space is just like space as I described it above: three-dimensional, with an objective notion of distance between points and angles between straight lines. But now, instead of imagining one big container in which things happen over time, I imagine that I have a *different* copy of space at each moment.

In his paper, Stein emphasized that the four-dimensional space-time structure I have just described—a structure sometimes called *Galilean space-time* in honor of Galileo Galilei—is *precisely* the structure needed for the physics of the *Principia*.[53] As we saw above, what Newton needed for the *Principia* was an objective, absolute, nonrelative sense in which a body is moving at *constant speed* in a straight line in space. A force, then, causes a change in speed or direction—that is, it causes a body to accelerate. That is what we have here: we can say that a body is moving at a constant speed, in a straight line, if the curve through space-time that you get by tracing the body's location at successive times is straight.

Stein's major insight was that this space-time structure is *not* what Newton *thought* he needed for the *Principia*. As I said above, when Newton defended absolute space and absolute place in the *Principia* and elsewhere, his reason was that there had to be a fact of the matter about whether a body occupied the same place at different times. That is, there had to be enough structure to say whether the body was at *rest* relative to absolute space. And the relationship between space and time that I have just described, the minimum one needs to assume to get Newton's laws of motion, does *not* let us say whether a body is at rest. The reason is that although space-time consists of copies of space stacked on top of one another, I have said nothing about how to link up places in one copy—space at one time—with the places in copies that come just before or just after. What I have given is a large collection of straight lines, cutting across the copies of space in all directions and corresponding to different states of straight-line, constant-speed motion. None of these straight lines in space-time is picked out as a special state, that is, as a state of rest. Instead, these lines give us many different, equally good ways of linking up places at one time with places at other times.

In other words, Newton was right that his laws required some fundamental, absolute notion of *acceleration*. But he was wrong that absolute acceleration required him to assume absolute velocity (or absolute speed). He did not need absolute space or absolute place after all.

To see the difference, think back to Newton's bucket example. When the bucket is rotating, the surface curves up toward the sides of the bucket. Newton took this to mean that there are physical consequences to a body's being in motion. But the bucket is *not* undergoing constant-speed, straight-line motion: instead, it is rotating. That means that each part of the bucket is constantly changing its direction of motion as it goes around in a circle. And this sort of change is a kind of acceleration. So what Newton's example really shows is that there are physical consequences to a body's accelerating.

You can contrast this sort of behavior with constant-speed, straight-line motion. Instead of a rotating bucket, imagine a bucket of water on the floor in a train. While the train is in the station, the surface of the water will be flat. But when the train starts moving, the surface will change: it will rise up toward the back of the train, and lower toward the front. This is a physical sign that the train is accelerating. But what happens when the train gets up to speed and continues on straight, smooth tracks without accelerating or decelerating? The surface of the water will once again be flat. The bucket will show no discernible difference from when the whole train was sitting at rest at the station.

All of this means that Newton's bucket *doesn't* show that there is a difference between absolute motion and absolute rest after all, as he and Clarke seemed to think. What it shows instead is that there is a physical difference between absolute acceleration—or at least, absolute *rotation*—and unaccelerated motion.

This is part of why Stein's paper was so startling. It showed that, by his own lights, Newton was wrong about absolute space—or at least, he was wrong that his own laws of motion *required* absolute space and absolute place. But does this mean Leibniz was right? Not exactly. As I said above, it is not clear that Leibniz ever developed a fully consistent theory of motion. But at least in the Leibniz-Clarke correspondence, Leibniz appears to claim that *all* motion is relative—including accelerated motions such as rotation. But if there is no fact of the matter about whether a body is accelerating, then there can be no fact of the matter about whether a body follows a straight line through space-time, or some sort of curved line. In other words, it seems Leibniz would have rejected the space-time geometry Stein describes. And without at least this much, Newton's laws do not even get off the ground.[54]

Stein was not the first person to characterize the Galilean space-time structure described above. That honor belongs to the mathematician Hermann Weyl, who introduced it in his book *Space-Time-Matter*, first published in 1918. Even so, Weyl was writing over a decade after relativity theory—the subject of the next chapter—had already forced physicists to reconsider the structure of space, time, and matter. This means that for the more than two centuries that Newton's ideas dominated physics, the relationship between space and time presupposed by Newton's laws was not fully understood. This fact is even more striking given that the dispute between Newton and Leibniz was well known, and at least in the century immediately following, the subject of ongoing controversy; for instance, both Voltaire and Emilie du Châtelet, two important eighteenth-century French commentators on the *Principia*, explicitly addressed Leibniz and Clarke's arguments on space and time in their commentaries on Newton. Likewise, Leonhard Euler, the most important mathematical physicist of the eighteenth century, wrote a paper in 1748 commenting on the correspondence (and siding with Clarke).[55]

And yet physics continued on, undaunted—and apparently unslowed—by this obscurity at its foundations. Often we have no choice, as physicists, but to press on, even when our best theories face apparently insurmountable conceptual and even mathematical problems.

Still, even if the details had not been fully worked out, some things were clear even in Newton and Leibniz's lifetimes. After all, they agreed on several things—and for the most part, these points of agreement were sufficient for physics to proceed. As we saw above, they agreed on the geometry of space: it is three-dimensional, with basic, fundamental relationships of straightness of lines, distances between locations, and angles between straight lines. They also agreed that at any instant, none of these geometrical facts depends on any choice of reference or observer. (They *disagreed*, meanwhile, about whether the same could be said about motion.) They also agreed—against Descartes and Aristotle—that space devoid of any matter is both conceivable and possible, though Leibniz did not believe the *actual* world could have any regions of empty space.

It was these points of agreement, plus a de facto assumption that whether absolute space really existed or not, some states of motion—constant speed, straight-line motion—had a special status that formed the bedrock of physics at the beginning of the twentieth century. All of these ideas were highly controversial before Newton's time and during his life. And while they had become dogma in the following centuries, *all* of them have since been revisited—and revised.

CHAPTER 2

Waves of Space Itself

IN THE SUMMER of 1900, a twenty-one-year-old Albert Einstein completed his undergraduate degree at the Zurich Polytechnic, in Zurich, Switzerland.[56] This was not the best or most prestigious college in Switzerland; it wasn't even the best in Zurich. Four students graduated that year in Einstein's program, which prepared students for teaching physics. Of them, Einstein's grades placed him dead last. He had likewise received the lowest score of the graduating students on his final research project, with the only lower score going to Mileva Maric, Einstein's girlfriend at the time and later his wife. (Maric's grades were so low that she was not permitted to graduate.)

Grades notwithstanding, Einstein was determined to earn a doctorate in physics. The natural first step, at least in turn-of-the-century Switzerland, was to find a professor of physics who would take him on as an assistant and support him as he wrote a doctoral dissertation. He began with the faculty at his undergraduate institution. Unfortunately, one of the two professors of physics at the Polytechnic had been so unimpressed by Einstein's performance that he had asked the university to give Einstein an official reprimand, for his severe lack of diligence; this same professor gave Einstein the lowest possible score—well below the threshold for passing—in his introductory experimental physics course. Einstein didn't stand a chance with him. So he wrote to the other professor, Heinrich Weber.

As it happened, Weber's opinion of Einstein was not much higher. He was the professor who had awarded Einstein the low grade on his final research project. Even though Einstein was the only graduate from the physics department to apply for an assistant's position that year, Weber rejected him.

He then wrote to the only other professor he'd interacted with: Adolf Hurwitz, who had taught him calculus. Again, no luck. Einstein had skipped most of his lectures, and Hurwitz had noticed. In the end, he did not receive any position at all—the only member of his class in physics or mathematics not to do so.

For two years, he worked odd jobs, mostly tutoring secondary-school students. He wrote to prominent physicists around Europe offering to be their assistant, but few even bothered to reject him. In November 1901, working on his own, he submitted a doctoral dissertation to the University of Zurich; it was rejected.[57] Three months later, Maric had a baby girl.[58] She and Einstein were not yet married, and she was living back at home, in Serbia, while he was still in Switzerland. Einstein didn't visit, and when Maric later joined him in Switzerland, she left the child back in Serbia. It seems Einstein never met his daughter. Finally, in June 1902, he was offered a position as a patent clerk in Bern, Switzerland. It was a far cry from the academic position he longed for, but he accepted it. He had little choice.

Einstein stayed at the patent office for seven years.[59] It was not a job designed to support research: he was expected to work six days a week. In 1903, he married Maric, and the next year they had a son. He continued tutoring on the side to support his growing family. Yet he managed to find time to pursue research in physics—largely on his own, though he would discuss his ideas with Maric and various friends in Bern and Zurich.

And then, in 1905, he revolutionized physics. Three times.[60]

In March of that year, he finished a paper in which he argued that for some purposes, light acts as if it is composed of what he called "light quanta"—discrete, particlelike packets of energy.[61] This was a major early step on the road to quantum physics; it was for this paper that he would win the Nobel Prize in 1921. It was his first revolution of the year.

The next month, he wrote his least significant paper of the year, in which he gave a new way of estimating the size of molecules.[62] He submitted it to the University of Zurich as a doctoral dissertation, and it was accepted, three and a half years after his first failed attempt. Eleven days later, he produced a related but far more ambitious paper on the statistical properties of small particles, under the hypothesis that they were being buffeted by tiny, unobservable atoms—a process known as Brownian motion.[63] This work would soon provide the theoretical basis for experiments demonstrating that unobservable atoms exist. Second revolution.

By the end of June he had written a fourth paper, "On the Electrodynamics of Moving Bodies."[64] It was here that he first presented the theory of relativity, the collection of ideas for which he is most famous. (It was in the year's fifth paper, developing these ideas further, that he presented his most famous *formula*, $E=mc^2$.[65]) Einstein's goal was to better understand how light moves, and he realized that, to solve this problem, he needed to revisit fundamental Newtonian concepts of length, simultaneity, and duration.[66] This led him to a radical new understanding of the structure of space and time—and ultimately, to a startling new picture of what the world could be like if there was nothing at all.

Revolution number three.

Despite Einstein's obscurity in 1905—and his unimpressive academic credentials—relativity theory was very quickly recognized as a major contribution.[67] Part of the reason it was so widely hailed was that it was a novel and compelling solution to a problem that everyone in physics recognized as important. This problem originated in the middle part of the nineteenth century, in the work of the Scottish physicist James Clerk Maxwell. Maxwell does not get the radio play that Einstein and Newton do, but on the list of greatest physicists in history, he would certainly make the top five.[68]

Maxwell made contributions to many subjects in physics, but the most important was his theory of electricity and magnetism.[69] Magnetic materials—known as *lodestones*—that could attract pieces of iron and some other metals were known to both the Ancient Greeks and the Chinese. Electricity was a more recent discovery, dating to around 1600, with experiments showing that rubbed amber would attract some substances, but not others. (The Ancients knew that rubbed amber had these properties, but apparently did not clearly distinguish electricity from magnetism.[70]) Electricity and magnetism were often studied together, and by the same people, because they were both mysterious forces that could apparently produce attraction and repulsion at a distance.

By the middle of the nineteenth century, when Maxwell began working on the subject, physicists understood that various electrical phenomena—static electricity, electrical currents, batteries—were all manifestations of the same underlying physics. The basic picture was that some bodies carry a special kind of property, known as *electric charge*. These charged bodies exert forces on one another, and these forces are the causes of electrical effects.[71] It was also known that electrical currents could be used to induce magnetic phenomena, suggesting that there was a complex interplay between electricity and magnetism. But beyond a few isolated mathematical formulas, there was nothing approaching a theory of electricity or magnetism—much less their relationship.

Maxwell changed that. He developed a detailed mathematical theory of electricity and magnetism, putting these forces on par with gravitation, as Newton theorized it in the *Principia*. The first step was realizing that electricity and magnetism are deeply linked. Maxwell's theory was not a theory of electric force or of magnetism. It was a theory of something new, which Maxwell called the *electromagnetic field*. Electricity and magnetism were just two different effects that this new sort of thing could produce.

Maxwell argued that what we had learned from the various known electric and magnetic phenomena was that space is filled with an "aetherial substance"—something that existed everywhere in the universe, permeating ordinary matter but also existing, in equal density, even in experimentally produced vacuums.[72] This aetherial substance—sometimes called the *luminiferous aether* or just the *aether*—was a kind of medium that could vibrate in various ways, and through which vibrations could move. It was these vibrations in the aether that made up the electromagnetic field.

To get a feeling for what this means, think about the air around you. Air consists of little particles of matter that can vibrate in various ways—and that can be made to vibrate by things like violin strings, vocal cords, or speakers. When this happens, the vibrations travel through the air, with each vibrating atom hitting nearby ones and passing the vibration along. They can also produce vibrations in other objects, such as eardrums or microphones, located some distance away from whatever created the vibration. This is how sound works: it is vibrations in a medium (the air) that travel over time.

The electromagnetic field, for Maxwell, was to the aether what sounds are to air: vibrations that are produced by electrically charged or magnetized objects, much as music might be produced by vibrating violin strings. These vibrations can then travel from wherever they originate to affect other things far away—by making other charged or magnetized objects move, the way music can make our eardrums vibrate.

The idea of an aetherial substance permeating all of space was not entirely new.[73] It bears many of the hallmarks of much earlier theories, such as those of Aristotle, Descartes, and Leibniz, who thought that all of space must be filled with some sort of plenum. But these theories had waned in the years since Newton, in part because Newton's theory of motion, unlike Aristotle's and Descartes', did not require a plenum for bodies to move, and in part because by the mid-nineteenth century, there were two centuries worth of experimental evidence that a vacuum could be produced in a laboratory by means of an air pump.[74]

Maxwell's theory of the aether changed this. Since the electromagnetic field consisted in vibrations *in the aether*, according to his theory, he took the strong experimental evidence that electricity and magnetism can pass through regions of allegedly "empty space," such as those created by air pumps, as evidence that these experiments had not really produced vacuums after all: there must still be aether even in chambers where all the air had been pumped out, since otherwise an electromagnetic field could not pass through them.[75] So, perhaps, despite the success of Newton's theory, empty space was not so empty after all.

Electricity and magnetism were not the only things Maxwell's theory unified. A crucial insight was that *light* is also a manifestation of the electromagnetic field. According to Maxwell, light is just a particular kind of vibration in the aether—namely, an instance of what is known as electromagnetic *radiation*, that is, a wavelike vibration in the aether that travels across space. (Other examples are radio waves, microwaves, and X-rays.) From a physical perspective, light is not all that special: it just happens to be the sort of electromagnetic radiation that our eyes are attuned to perceive, much as sound is just the sort of vibrations in the air that our ears are attuned to perceive.

Still, the fact that we *can* perceive light makes it special to us, and the behavior of light, in particular, is what attracted Einstein's attention.[76] The reason was that the motion of light and other electromagnetic radiation led to conceptual problems for Maxwell's theory. The puzzle concerned how to interpret the *speed* of light, a parameter that was baked into Maxwell's equations governing the electromagnetic field.

Recall from the previous chapter that, according to Newton's physics, whenever we measure the speed of another object, we are measuring a *relative* quantity: how fast the other object is moving relative to us, or to our measuring apparatus. A speedometer or a radar gun only tells us the speed of a car relative to the road. Likewise, physicists reasoned in the nineteenth century, when we measure the speed of light, we are making a relative measurement: we are measuring the speed of light relative to us.

Such measurements had been made, and it was known even in

Newton's time that light travels at a finite speed.[77] By the middle of the nineteenth century, when Maxwell began writing on the subject, this speed had been measured fairly accurately, and was known to be about 300,000 km/s. This experimental value was used in Maxwell's theory, to great success.

This is where the puzzle arises. Nothing is wrong with plugging an experimentally determined value into a new theory to get predictions. That happens all the time in physics. But there was something strange about a theory that took, as a fundamental constant of nature, the speed of light *relative to observers on the surface of the earth*. After all, the implication would be that a basic law of nature governing the electromagnetic field was specially tuned to require a number that only we, on the surface of the earth, could determine. What makes us so special? What if we went to the moon, or to Mars, or even just up to the top of a mountain, where our motion through space would differ from our motion at sea level, and tried to measure the speed of light from there? Presumably, given everything nineteenth-century physicists knew about measuring speeds, we would get a different answer, reflecting the fact that our new measuring apparatus was itself moving relative to our old measuring apparatus.

If the speed of light is to be a fundamental constant, physicists reasoned, there must be some special perspective that yielded its real value. Of course, Newton believed that there *was* a special perspective of just this sort: that of absolute space. Arguably, it would make sense for speed relative to absolute space—absolute speed—to play a special role in our fundamental theories of nature: it would no longer be *us* appearing in the laws of physics but space itself.

Maxwell and his contemporaries assumed that, although the experimentally determined speed of light seemed to work well enough in his equations, the real value that the equations called for was something like the absolute speed of light—not necessarily its speed relative to absolute space, but at least its speed *relative to the aether*. In other words, if, as Maxwell believed, the electromagnetic field consisted in vibrations in the aether, and the aether was some sort of substance spread out through all of space, then the aether must determine a special state of motion. It would be speeds relative to this aether that would reflect the *real* speed of light. The aether would end up playing a role in physics very similar to the role Newton imagined for absolute space.

All of this made perfect sense, in theory. The trouble was that if any of it was right, there would presumably be some experimental consequences. For instance, no one believed the earth was at rest relative to the aether. The earth is rotating around its axis, and also orbiting the sun, and presumably it would be an enormous coincidence if the aether were *also* moving in precisely these ways. Nineteenth-century physicists quite reasonably thought we should be able to detect effects resulting from the earth's motion relative to the aether. Likewise, if we measured the speed of light relative to bodies moving on the surface of the earth—say, relative to different sensors that are moving relative to one another—we should detect differences in the speed of light.

These experiments were performed.[78] And they found nothing. No evidence of motion of the earth relative to the aether, and no differences in the speed of light as measured by moving bodies. Nothing. Fundamental Newtonian ideas about space, time, and motion seemed to imply that light must have different speeds when measured by moving apparatuses, but the best experiments anyone could do showed that it did not.

This was the puzzle that Einstein took up in his papers on relativity theory in 1905. He offered a deceptively simple solution:

the speed of light, he claimed, is actually the *same* for every non-accelerating observer, no matter how fast these observers are moving relative to one another. There is no special state of rest relative to the aether: any state of (unaccelerated) motion is just as good as any other, and all of them will yield the *same* results for measurements of the speed of light. There was thus no mystery about why Maxwell's theory worked so well with the experimentally determined speed of light, even though these experiments were not performed by observers at rest relative to the aether. The experimentally determined speed of light was the speed of light, full stop.

This idea, that the speed of light is the same for every observer, should strike you as very strange. In fact, it flies in the face of everything we know about relative motion, from Newtonian physics and everyday life. To borrow some classic examples from Einstein himself, suppose that you are riding on a train, traveling at more or less constant speed. You throw a baseball toward the front of the train. You could measure the speed of the ball using whatever instruments you like on the train and you get some answer.

Now suppose that you throw the ball just as the train happens to be passing a station, and your friend, sitting on the station platform, *also* measures the speed of the ball. Of course, the friend will get a different answer: as measured from the platform, the ball will be moving faster, because the train is also moving in the same direction. This is just a fact of life that figures into all sorts of ordinary activities, from changing lanes on the highway to catching a Frisbee on the run.

But according to Einstein, while this basic fact of life may apply, at least approximately, to some motions, it doesn't apply to the motion of light. If, instead of throwing a baseball, I turn on a flashlight and measure the speed of the beam as it approaches the front of the train, I would get about 300,000 km/s. If my friend on the platform measures the speed of the same beam, she will get 300,000 km/s—just as if she, too, had been at rest relative to the source of the light. And this will hold no matter how fast the train is moving.

I said above that Einstein's solution to the problems facing Maxwell's theory was deceptively simple. The reason it is deceptive is that, if there is anything in the world that moves the way Einstein argued light does, then "motion" cannot mean what we thought it meant. Einstein's claims about the motion of light were flat-out inconsistent with the structure of space, time, and motion as described by Newton and accepted by generations of physicists.

Einstein understood this, and he began his first paper on relativity theory by redefining these notions. He started with a question: What do we need to assume to measure the motion of a body?[79]

Imagine you are standing on a football field, next to a talented quarterback throwing a football down the field. How might you measure the speed of the ball? One way would be to look at your watch the instant the ball leaves the quarterback's hand, say at the twenty-yard line, and record the time. You can then keep checking your watch at regular intervals, say when the ball passes the thirty-yard line, the forty, and so on, recording the time at each interval.

From these measurements, you can work out how much time it takes the ball to travel ten yards, and then infer what its speed must be. But this inference relies on an assumption: namely, that the instant at which you see the ball cross, say, the thirty-yard line (and the instant you look at your watch) is *simultaneous* with the instant at which the ball *does* cross the thirty-yard line. But strictly speaking, this isn't true. In order for you to see the ball cross the thirty-yard line, light needs to reflect off the ball and then travel to your eye. Thus the time we record is *not* the time at which the ball crossed the thirty-yard line, but rather a few instants later, when the light reaches you. How much later? Precisely the amount of time it takes light to travel ten yards, from the thirty-yard line back to where you stand at the twenty-yard line. So this procedure requires you to know the speed of light to work out which instant—as determined by your watch—was actually simultaneous with a distant event—that is, the ball passing the thirty-yard line.

Let's go back to the drawing board. Here's another procedure that seems to avoid the issue just noted. Instead of doing all of the recording from the twenty-yard line, you could enlist a friend to go and stand at the thirty-yard line and make that measurement for you. Then you record the time at which the ball leaves the quarterback's hand, and your friend records the time at which the ball passes her.

Now we don't have to worry about the time it takes for light to travel back to you from the thirty-yard line. But there is a new problem. These two different measurements can only be used to calculate the speed if you and your friend know that your watches are synchronized: that is, that at any instant, both of them show the same time. And how can we tell if that is the case? Even if you start with identical watches, and you synchronize them while you're standing next to one another before the ball is thrown, there is still a chance that they have come out of synchrony by the time the measurements take place. Perhaps your friend jostled her watch as she ran out to the thirty-yard line, or perhaps your watch battery is running low.

It might seem that there's an obvious solution. To check that your watches are still synchronized when everyone is in position and it is time for the ball to be thrown, you and your friend can signal to one another to see if your watches still line up. But those signals will take some time to pass between you, so to work out if the watches are synchronized, you need to account for that travel time.

You can begin to see the pattern. In *both* of these measurement procedures, you need to come up with a way of determining when two events, occurring some distance apart, are simultaneous. And these determinations are difficult: you need to find a way to signal, and you need to figure out how fast the signal moves—which, in turn, requires you to figure out how simultaneity works, leading to a regress. One way out is to simply assume that everyone agrees on which events are simultaneous, that our clocks are perfectly reliable, and so on. This, in effect, is the Newtonian approach—and also the approach we take when we are trying to catch buses and meet friends for drinks. With this assumption, however, the speed of light cannot be the same relative to all observers—for the same reason that the speed of a car cannot be the same relative to all observers.

But what if, instead of making this assumption, we adopted Einstein's assumption that the speed of light *is* the same for all observers? Then we have a different way out of the quandary: we can use light, or other electromagnetic radiation, to send our coordinating signals. (The speed of light, meanwhile, becomes easy to measure: just bounce a beam off of a mirror some known distance away, and figure out how long it takes the light to travel there and back.) Since we all agree on the speed of light, no matter what our state of motion, we can easily do the necessary calculations to account for travel time in our measurements.

Adopting this viewpoint may seem innocuous enough—even pragmatic, since it provides a sort of engineering solution to the problem of how to measure motion. But it quickly yields a startling conclusion: if this is how we determine simultaneity, then different observers, moving relative to one another, will make *different* determinations of when two events are simultaneous. In other words, simultaneity—what is happening in the universe *right now*—must become a relative notion, just like motion.

None of this—the nonrelativity of the speed of light or the relativity of simultaneity—can be accommodated by the Galilean space-time structure of Newtonian physics that I described in the previous chapter. To make sense of it, we need a completely new picture of the geometry of space-time.

Einstein knew that relativity theory amounted to a redefinition of space and time. But he did not recognize that his new theory amounted to a new understanding of space-time geometry.[80] That insight was due to Hermann Minkowski, one of Einstein's former mathematics professors at the Zurich Polytechnic. He went on to describe a new kind of space-time geometry, now known as Minkowski space-time.

Physicists today take for granted that Minkowski space-time is the right geometrical setting for special relativity. But Einstein, when he first heard of it, was unimpressed. He dismissed Minkowski's work as "superfluous learnedness" and insisted (perhaps jokingly) that he didn't understand it.[81] (Minkowski, for his part, shared his colleagues' opinion of Einstein, writing to a friend that "in his student days Einstein had been a lazy dog.") Over the next six or seven years, however, Einstein's views changed dramatically. Eventually he realized that Minkowski had uncovered something very deep about relativity theory.

Like Galilean space-time, Minkowski space-time is a four-dimensional "space" of events in space and time.[82] Just as in classical physics, any two points, p and q are connected by a unique arrow originating at p and ending at q, and likewise a unique arrow originating at q and ending at p (the opposite of the first arrow). Once again, this means there is a notion of "straight line" baked into the space-time structure.

But this is where the similarity stops. In particular, there is *not* a relation of "simultaneity" between events. That is, given two events, it is no longer the case that they either occur at the same time or they do not. Space-time simply does not have the structure to make such determinations. Without this relation, we cannot carve up space-time into slices of three-dimensional space at different times. And without this slicing, it does not even make sense to ask how far apart two simultaneous points are in space. After all, there is no "simultaneity," and there is no "space." Nor does it make sense to ask about the duration between events, again because the structure presupposed by our earlier notion of "duration" has been thrown out the window.

It goes without saying that this is a lot to give up. Much of our ordinary thinking and talking about space and time presupposes a classical space-time structure—the structure necessary to say, "I brushed my teeth while my kids were eating breakfast downstairs" or "I woke up just as the alarm went off." And yet if we accept that relativity theory provides a more accurate representation of the world than Newtonian physics—and the empirical evidence overwhelmingly suggests that it does—it seems we are forced to give up just this structure.[83]

This does *not* mean we have to give up our ordinary ways of speaking and thinking. After all, these serve us perfectly well for all the purposes we care about. But we have to recognize that ordinary talk of simultaneity, space, and time is at best a rough-and-ready approximation of what is going on, and our intuitions can take us only so far toward developing a more systematic understanding of the world.

So much for what we have to give up to get from Galilean space-time to Minkowski space-time. What sort of structure do we have instead? In Minkowski space-time, we have one basic relation known as the *space-time metric*. The space-time metric allows us to classify directions in space-time—arrows at a point—into three categories: *timelike*, *spacelike*, and an intermediate whose significance I explain presently, *lightlike*. It also allows us to determine a quantity known as the *space-time distance* between different events.

Space-time distance is, as the term suggests, a distance between events in space and time. But when you are trying to figure out how to get to the grocery store in a new town, the space-time distance between events isn't what you're after. You want our ordinary notions of duration in time and distance in space. The space-time metric encodes this information, too, but extracting it requires a little work.

We often think of ourselves—or any physical body—as moving through time, from the past to the future. This idea is captured in Galilean space-time by representing bodies with curves that intersect different slices of space. In Minkowski space-time, the idea is captured using curves moving in a timelike direction, or *timelike curves*. These are possible trajectories through space-time for massive bodies or for idealized observers. If two events are connected by a timelike curve, then those events happen at different times, one earlier than the other. Everyone agrees on this.

Ordinary people, represented by timelike curves in Minkowski space-time, can make measurements of duration as they go about their business. They do this in the usual way, by using clocks. These measurements correspond to the length of the timelike curve they follow, as determined by the space-time metric. In Galilean space-time, everyone agrees on the duration between events. There is an unambiguous amount of time that passes between one instant and any later instant. But in relativity theory, that isn't true. Duration is something measured by an observer, by a clock that the observer takes along their trajectory through space-time. Identical clocks, following different trajectories between the same two events, will record different durations.

Straight lightlike lines, meanwhile, represent very special trajectories: the possible trajectories for light. These can be thought of as a boundary for the timelike curves, representing a maximum speed limit through space-time. They form a cone structure at each point, with all of the timelike curves falling inside the cone and the spacelike ones outside it. These special trajectories for light are built into the structure of space-time, which explains why all observers, whatever their state of motion, agree on how fast light moves. Light just moves along a different kind of curve in space-time than you or I.

I noted earlier that the fact that light has the same speed for all observers makes it a potent tool for synchronizing clocks. But it is equally useful for other things, such as measuring distances. In Minkowski space-time, these procedures are represented using lightlike lines.

Let me spell this out more concretely. Suppose you want to measure the distance between yourself and some other object. You can do this with a flashlight and a wristwatch: just affix a mirror to the thing whose distance away you'd like to measure, and then shine the flashlight at the mirror. By measuring the round-trip transit time, you can infer how far the light had to travel. All of this is represented easily in Minkowski space-time. You and the mirror are both represented by timelike lines, whereas the light follows a lightlike line emanating from the point in space-time where you turn on your flashlight. This lightlike line intersects the mirror's trajectory and then bounces back, intersecting your trajectory again later on. The duration, as determined by the space-time metric, between when you turn on the flashlight and when the light's trajectory intersects yours again is the time it takes for the light to travel to the mirror and back.[84]

Using the same basic construction, you can also figure out when the light reached the mirror: namely, at the time halfway between whenever you turned on your flashlight and when the light returned after bouncing off the mirror. This is how determinations of simultaneity work in Minkowski space-time. And by collecting up all of the mutually simultaneous events in Minkowski space-time, using basically this procedure, you can get a notion of *space* at an instant, as determined by some observer. And the distance between you and the mirror that you determine by the procedure described above is precisely the space-time distance between the event where the light hit the mirror and the event in your own history that you determine is simultaneous with it. This is how the space-time metric encodes what we ordinarily think of as distance—that is, distance in space.

These procedures let you carve up Minkowski space-time into space and time, much as in Galilean space-time. But the crucial difference is that the results of this construction depend on the state of motion of the observer—that is, on the timelike curve with which one begins. If we began with a *different* observer, represented by a different timelike line, that observer, using the same procedure, would get different answers for lengths and for simultaneity. Some distances would seem longer, others shorter; simultaneous events would happen at different times, and events at different times would become simultaneous.

Of course, this is just what we should have expected, given what

I said earlier about the relativity of simultaneity. Now, though, we see that the surprising features of relativity theory all come down to how our own trajectories through space-time intersect the paths of light rays and what inferences about simultaneity and distance we draw from that.

Although Einstein was not (yet) thinking geometrically, the substance of the revisions to simultaneity, length, and duration I've described all appeared in the original 1905 paper. It was with these revisions that he solved the problems in Maxwell's theory: once we stipulate that everyone agrees on the speed of light, we no longer worry about its motion relative to the aether. But Einstein went on to make a more radical claim: *there is no aether*. As far as Einstein was concerned, the role the aether played in Maxwell's theory was to make sense of the speed of light. Without that role to play, the aether was unnecessary.[85]

But where does this leave Maxwell's theory? After all, Maxwell believed the electromagnetic field consisted in vibrations in the aether. If there is no aether, then what could be vibrating? The answer is that the electromagnetic field *itself* is a new kind of stuff. What used to be conceived as different oscillations in the aether are actually different configurations of this new sort of entity—the field. These configurations can be wavelike or oscillatory, but they are not oscillations *in* anything, as the aether picture suggested. The aether turned out to be just a stepping-stone to reckoning with the idea that, all around us, there is an electromagnetic field.

I said earlier that the aether was a bit like the plenum that Aristotle, Descartes, and Leibniz all believed filled space, which meant that, in a world where electromagnetic phenomena are possible, empty space is *not* possible, since an aether must be present everywhere. There could not be *nothing*—or at least, there could not be regions of space where there is nothing as long as electromagnetic phenomena could occur.

Einstein rejected the aether just as Newton rejected the plenum. But in its place, we are left with the electromagnetic field, which *also* pervades all of space. All indications are that, in fact, the whole universe has a steady, low level of electromagnetic radiation pulsing through it, known as the Cosmic Microwave Background.[86] So in a sense, we have left the aether behind but replaced it with a different kind of "aetherial substance" that also pervades all of space, at least in our universe.

Still, we might wonder whether, even if our universe *does* have an electromagnetic field filling all of space and time, it is at least *possible* to have a universe with no electromagnetic field—or even a region of space and time with no electromagnetic field.

This question turns out to be more difficult than it seems. We measure the electromagnetic field by a quantity known as *field strength*. This quantity oscillates when radiation passes through a region; charged bodies produce field strength, and the larger the field strength, the more acceleration charged bodies will experience. Roughly, we can think of the field strength as telling us how much electromagnetic field is present somewhere.

Now suppose there is some region of space-time where the field strength is zero. This means, for instance, that there is no electromagnetic radiation in the region—no light, no X-rays, and so on. But does it mean that there is no electromagnetic field at all? On one way of thinking, the answer is certainly "yes." If the field strength is zero, there would be no detectable electromagnetic effects. The "amount" of field present would be zero, which is just another way of saying there is no field present.

But there's another possibility. One might think that the electromagnetic field fills all of space, but that it may be configured in different ways, depending, for instance, on the distribution of charges in the universe and other factors. The field strength is just a way of describing how the field is configured. And one possible configuration—no different, in principle, from any other—is the one in which the field strength is zero everywhere in some region. On this way of thinking, saying the field strength is zero is just the same as saying the field isn't doing anything—but not the same as saying there is no field at all.

Think about it this way: Imagine a baseball stadium filled with people. These people might do "the wave," that is, stand up with their arms in the air in a regular pattern that looks like it's propagating through the stadium. If they do this, I could describe their configuration in a region of the stadium by telling you how many people are standing up in that area. This would be a kind of "wave strength." Sometimes, though, there would be no wave—that is, no one standing. The wave strength would be zero. But this doesn't mean the stands are empty. Similarly, no field strength need not mean no electromagnetic field.

Which of these is right? In classical electromagnetic theory, as I have described it here, arguments could be given for either view. (I happen to prefer the second.[87]) But as we will see in the next chapter, moving to quantum field theory seems to provide evidence for the second way of thinking about the field. This will turn out to have real significance for what it means to say that there is "nothing" in a region of space and time. Nothing is not the *absence* of stuff; instead, it is just one possible *configuration* of stuff.

When Einstein's 1905 papers were published, he was still a patent clerk in Bern.[88] He hoped the flurry of results—which he rightly believed to be extraordinary—would finally lead to academic recognition. And to some extent, they did. Important and influential European physicists began to publish on relativity theory.

But appreciative citations and positive attention didn't put food on the table, and no job offers appeared. In 1907, he tried applying for an academic job again, this time at the University of Bern. The position he applied for was neither prestigious nor lucrative. It was an entry-level position, appropriate for someone who had just finished a PhD. One member of the hiring committee supported Einstein's application, but he couldn't sway his colleagues. The position went to someone else.

Einstein began to despair of ever finding an academic job. If his accomplishments of 1905 weren't enough, what could be? He began working on other projects, including designing and trying to sell electrical devices.[89] He also applied—unsuccessfully—for jobs in secondary schools, teaching physics and mathematics to teenagers.

The next year, he tried again for a position at the University of Bern. At last he was appointed. But it was a Pyrrhic victory: he was paid so little that he could not afford to quit his job at the patent office, which meant the longed-for post was little more than another time suck, standing in the way of his research. (He did, at least, get to use the university library: before that he had little idea what others had done to expand on his theory of relativity, because he had no access to the current literature![90])

Once his foot was back in the door, however, things improved quickly. In early 1909, he was offered a position as an associate professor back at the University of Zurich. Though it was better than the Bern position, and Einstein ultimately accepted it, the Zurich position was still not an attractive post: in the German system, including Switzerland, an associate professor typically works under the supervision of a full professor, often as a kind of glorified research and teaching assistant.

Two years later, in January 1911, Einstein was offered a proper professorship, at the University of Prague. He accepted it—only to accept yet another professorship in 1912, back in Zurich.

This new appointment was not at the University of Zurich, where he had done his PhD and then briefly served as associate professor. It was at the Polytechnic, where he had been an undergraduate. And so, a little over a decade after he left, Einstein found himself back where he began. Now, though, things were different. The beleaguered and mediocre student was long forgotten: Einstein returned as a conquering hero, the local boy made good. And in the meantime, the Polytechnic itself had been transformed: in 1911 it was made a full university, with the right to grant graduate degrees; it was also renamed the Swiss Federal Institute of Technology.

Coincidentally, Einstein wasn't the only member of his graduating class to hold a professorship at the newly minted university. An old friend, Marcel Grossmann, held a position there, too, as professor of mathematics.[91] Grossmann had been the star mathematics student in Einstein's undergraduate cohort, receiving perfect scores on exams that Einstein barely passed. Indeed, Grossmann had helped Einstein get through the courses by sharing his notes on the lectures. (This is not to say Einstein *struggled* with the mathematics, exactly. He skipped the lectures, and Grossman covered for him.) Later, it was Grossmann who helped Einstein secure the job at the patent office.

When Einstein returned to Zurich in 1912, he quickly sought out Grossmann. It was just like the old days: he needed some help with math.

In the years since 1905, Einstein had come to think that the theory of relativity was incomplete. From a modern perspective, the heart of the matter was an incompatibility with Newton's theory of gravitation. According to Newton's theory, bodies exert gravitational force on one another across long distances instantaneously. But the very notion of "instantaneous" action is incompatible with relativity theory, since there is no fact of the matter about when two events are simultaneous. So Einstein needed to find a new theory of gravitation—one that was compatible with the space-time structure of relativity theory. (It is interesting to note, though not important for our purposes, that Einstein understood the motivations for the theory somewhat differently while he was working on it.[92])

He worked on this project on and off from 1907, but the theory was still inchoate when he knocked on Grossman's door. He had come up with a handful of guiding insights. For instance, he thought gravity should influence the trajectories of light beams. As we saw above, light follows special paths through space-time, represented by straight, lightlike lines. What Einstein realized was that, in the presence of gravity, those lines should not be straight any longer. They should be curved, with the amount of curvature depending on the gravitational influence of other matter in the universe.

But Einstein was struggling to find a theory in which light would do this—and to relate this behavior of light to other kinds of gravitational interactions in a way that would be consistent with the rest of relativity theory. So he turned to his friend. Einstein described what he had come up with so far—some guiding principles, sketches of special cases where he had a proto-theory that seemed to work—and asked if Grossmann had any ideas about what sorts of mathematics Einstein should learn if he wanted to develop the theory further.

Grossmann left the meeting excited by what Einstein said, but unsure of what to suggest. After giving it some thought, he had an idea. Einstein needed more geometry.[93]

The two space-time structures I have described so far—the Galilean space-time of Newtonian physics, and the Minkowski space-time of relativity theory—seem to present divergent pictures of the structure of space and time. But there is one way in which they are exactly the same: in both of these space-time structures, any two points can be connected, in a unique way, by an arrow, or straight line, that begins at one of the points and ends at the other. In the language of modern geometry, this means that both of these space-time structures are *flat*.[94]

I am using the word "flat" in a mathematical sense, but it is very closely related to ordinary usage. When mathematicians say that a space is flat, they mean it is not *curved*. Curved spaces do not have the property that their points can be connected uniquely by arrows. To see this, just think about a basketball. The surface of a basketball—or the hood of a car, or a horse saddle, or the line of your chin—is curved. If I took a straight line, such as the edge of a ruler, and put one end on the surface of the basketball, the *other* end wouldn't touch the basketball at all. It would run right off the surface. This means that straight lines starting at one point of the basketball *don't* connect that point to other points of the basketball; conversely, no line that you could draw on a basketball connecting different points would look straight.

Now, there are some flat spaces where I cannot do this either. For instance, a piece of paper lying on a table is (to a great approximation) flat. And usually, I could put a ruler on the paper and connect any two points with the edge of the ruler. But what if

I cut out a heart from the middle of the page? Then I couldn't draw a straight line connecting points on either side of the heart, even though the paper is still flat. But what *is* true is that if I can connect any two points with a unique straight line, with all of the properties of arrows that I have described, then that space must be flat.

The fact that I can't connect points on a basketball with the edge of a ruler doesn't mean that I can't draw lines on a basketball connecting different points. It also doesn't mean that I can't draw special lines that are, roughly speaking, *as straight as possible*. For instance, if I pick any two points on the surface, I can imagine drawing a line that wiggles around a lot on the way from the first point to the second point. And I can imagine drawing lines that wiggle less and less, until I get a line that doesn't wiggle at all. This line isn't straight in the strictest sense, since there are *no* straight lines on the surface of a basketball. But it is straighter than other lines.

Recall that in Minkowski space-time, light moves along straight lines that have a special status determined by the space-time metric. What Einstein and Grossmann soon realized was that if gravitation bends the trajectories of light, then gravitation must be changing the curves that light can travel along. They went on to infer that this meant that gravity was actually changing the geometry of space and time: light, instead of traveling along straight, lightlike lines in Minkowski space-time, must be traveling along *straightest* lightlike lines in a curved space-time. Similarly, other small bodies, instead of moving along straight timelike lines in flat space-time, are moving along straightest possible timelike lines in curved space-time. Hence the idea of space-time curvature was born.[95]

In the theory that Einstein ultimately developed—known as

general relativity, in contrast to the 1905 theory, which is now called *special relativity*—this curvature of space-time, manifested physically in the motion of small bodies and light rays, replaces what we used to think of as gravitational force. According to Newton's theory, massive objects accelerate toward one another because they exert a gravitational force on one another. In Einstein's theory, these same objects are not accelerating at all: they are both moving along the straightest lines they can, in a space-time in which *no* lines are truly straight.[96]

The curvature of space-time, meanwhile, depends on how these very same bodies are distributed through space and time. In other words, matter actually changes the geometry of space-time, so that those straightest lines move toward one another, in much the same way that two lines on the surface of the earth, beginning at the equator and pointing due north, gradually approach one another as they head toward the North Pole. This means that massive bodies tend to attract one another—or act as if they do—just as we would expect from Newtonian gravitational theory. But there is no force. What we used to think of as gravitation is just a manifestation of this space-time geometry.

One of the most important changes wrought by general relativity is the idea that the geometrical structure of space and time in fact depends on the stuff in space and time. It is this stuff that determines the curvature of space-time—and in turn, determines how bodies move. But what happens if there *isn't* any stuff in space-time? What is the structure of empty space-time in general relativity?

One possibility is that if space-time were empty, it would simply look like Minkowski space-time. In other words, if stuff produces curvature, and there is no stuff, you might assume that there would be no curvature. And Minkowski space-time is a perfectly natural candidate for a space-time that has no curvature, and no stuff.

But soon after Einstein discovered general relativity, in 1915, other physicists noticed something. The central equation of the theory, which we now call Einstein's equation,[97] relates properties of matter—namely, the energy and momentum of matter—to the curvature of space-time. *Solutions* to this equation represent possible universes according to the theory. They are four-dimensional space-times whose curvature and energy and momentum are coordinated as required by Einstein's equation. What these physicists realized, however, is that simply specifying how matter is distributed in space-time is not generally enough to fix the curvature of space-time. In fact, there are infinitely many possible universes (or better, universe-histories, since we are speaking of four-dimensional space-times) allowed by general relativity in which there is no *stuff* at all. These are all solutions to Einstein's equation that differ from one another in their curvature—even though none of them contain any matter.[98]

In other words, although the distribution of stuff is related to curvature, no stuff does *not* guarantee no curvature. This means that empty space in general relativity is far, far richer than in special relativity—or in Newtonian physics. Indeed, it is so rich that it stretches the meaning of "empty" nearly to the breaking point.

The most striking way to make this point is to describe an example of an empty universe that was first discovered in 1915, the same year that Einstein discovered the fundamental equation of general relativity. This example was found by the German physicist Karl Schwarzschild, and for this reason it is known as Schwarzschild space-time.[99] Schwarzschild's goal was to use general relativity to describe the curvature of a space-time in which there is nothing but a perfectly spherical object, like a beach ball or (approximately) a star.

Schwarzschild approached the problem by assuming there was some object—the beach ball—and that he was going to describe space-time *outside* of it, where there wasn't any other stuff. He did not aim to say anything about the beach ball itself, or about the curvature of space-time inside the beach ball. Pragmatically, this made good sense. For instance, if you want to understand the solar system, you might initially try to set aside questions about what happens inside the sun and just try to describe the motion of planets located far away from the sun.

Schwarzschild successfully solved the problem he set for himself. But he ended up finding more than he bargained for. It turned out that the solution he came up with was perfectly valid even if there was *no beach ball at all*. In other words, the solution made sense, insofar as it appeared to describe a possible physical situation, even if there was no matter located anywhere in the universe.

The important thing about this example is that, even without matter, Schwarzschild space-time is not the same as Minkowski space-time. It is not flat. In fact, if you imagined putting a small object in this otherwise empty space-time (or an observer in a rocket ship) to study its properties, you would find that the object would behave very differently from how it would behave in Minkowski space-time. In Schwarzschild space-time, the object would move much as we would expect something to move in a universe with a large, massive object, such as a star. It would act as if it were falling *toward* something. But there is nothing—at least, no matter of the ordinary sort—with which the object would ever collide.

Instead, something weirder would happen. Imagine that instead of some inanimate object, you have a friend (or better, someone you don't like very much) in a spaceship in Schwarzschild space-time, and you are also in a spaceship some long distance away. Suppose that as your friend "falls" away from you, toward the mysterious nonthing that by all appearances is pulling her in, she periodically sends you messages with light rays, updating you on her status—including the time at which the message is sent, according to your friend's watch. You keep reading the messages, but as time passes, they appear to arrive more and more infrequently, even though she reports that she is sending them regularly—say, every five minutes. Ultimately, they stop arriving altogether.

Why do the messages stop? The reason is that if your friend "falls" long enough, she will eventually pass into a region of Schwarzschild space-time where the curvature of space-time is so extreme that light rays cannot escape: instead, they curve back on themselves, so that even light behaves as if it is falling, much as your friend is.

The threshold of this region of Schwarzschild space-time has a name: it is known as an *event horizon*. It is a point of no return: once you pass the event horizon, you can never pass back out, no matter how much rocket fuel you use. You cannot even send a signal out, because light cannot escape either. Your friend will just keep falling until eventually she simply pops out of existence.

You have probably heard of event horizons before—or at least you have heard of the expression "black hole," which is a colorful (or not so much) name for the region of space-time enclosed by the event horizon.[100] In effect, what Schwarzschild discovered, without entirely realizing it, was that Einstein's theory of relativity predicts the existence of black holes.

It is tempting—and quite common—to think of black holes as *things*. They are often described as the remnants of stars that have collapsed under their own weight. They have locations, roughly speaking; they influence the behavior of other objects in much the same way as stars and planets; and there is even a sense in which they can be said to have a mass, just as you or the moon or any other object does.

From this point of view, it is natural to describe Schwarzschild space-time as a universe containing *something*—namely, a black hole. But we have to be very careful. Even though Schwarzschild space-time has a black hole, there is not actually any *stuff*—that is, any matter—in it, anywhere or at any time. Not a speck of dust. If you asked any observer riding a rocket ship around in Schwarzschild space-time, "Is there any matter near you?" that observer, looking around, would report he or she doesn't see anything at all. It is a solution to Einstein's equation that is every bit as empty as Minkowski space-time.

For this reason, the black hole in Schwarzschild space-time is different from the black holes that we believe exist in our own universe. Those, we think, formed when massive bodies such as stars collapsed under their own weight. Schwarzschild space-time, on the other hand, describes a universe in which there has *always* been a black hole, eternal and unchanging. It is not the remnant of anything. In a sense, this is why it is such an important example. It shows that a black hole is a purely geometrical phenomenon, possible even in a universe where there was never any matter.

So is Schwarzschild space-time empty, or not? A clean "yes" or "no" seems impossible. There is no matter located anywhere in the space-time. Yet such a universe would have all the hallmarks of a universe containing some large, massive body, like a super-dense star. Certainly, if a friend told you that a room was empty, and you walked in to find a black hole, you'd be rightly annoyed—not that you'd be able to do much about it—even though in some strict sense, your friend would have been right. This is what I meant when I said that general relativity strains the meaning of "empty" to a breaking point. "Nothing but a black hole" sure seems like a lot.

Schwarzschild space-time was the first solution to Einstein's equation that anyone discovered (other than Minkowski space-time). Others soon followed. An important example was due to Willem de Sitter, a Dutch mathematical physicist. Like Schwarzschild space-time, de Sitter space-time is empty, in the sense that no observer located anywhere in space or time would ever report seeing any matter. Also like Schwarzschild space-time, de Sitter space-time is curved, and so the physics of de Sitter space-time is very different from that of Minkowski space-time.[101]

The details of de Sitter space-time do not matter all that much. What *does* matter is that, when de Sitter first described his solution to Einstein in a letter in 1917, comparing his solution to one Einstein had recently proposed, Einstein *hated* it. And what he hated most was just the feature that makes it most interesting for us. It was an example of a space-time with many of the features one might expect the real universe to have, but in which there is not any stuff at all.

The reason Einstein hated it was that he believed general relativity was—or at least, should have been—a theory in which space-time geometry was *fully* determined by the ordinary matter spread throughout space and time. Indeed, one of the reasons he rejected Newtonian physics was that Newton's theory presupposed all sorts of structure, independent of any of the stuff that actually exists in and moves through space and time. Einstein thought this sort of structure was metaphysical mumbo-jumbo, and he wanted to get rid of it.

The theory he produced, however, was in many ways worse.

While there was no longer a fixed background space-time structure, there was now an enormous amount of freedom in just what the space-time geometry is, even after the matter distribution is fixed—and indeed, as we have seen, even if there is no matter at all.

In other words, Einstein didn't like Newton's theory because it simply assumed certain default trajectories along which bodies would move. If asked, "Why does a body go this way rather than that?" Newton could say nothing more than, "That's just how space-time is." Einstein wanted a better answer to this question. He wanted to be able to add, "And space-time is the way it is because the stuff in the universe is arranged in such and such a way." But it turned out he could say no such thing, since there are many ways space-time could be, even without any matter at all.

Over the subsequent decades, Einstein and many others came to understand that the de Sitter and Schwarzschild space-times were just the tip of the iceberg. In fact, there are infinitely many ways that an empty universe could be—despite Einstein's strong preferences to the contrary. This shows just how little control we may have over the physical theories we create: in many cases, even as we build theories with certain goals in mind, the mathematics—and the world—just doesn't cooperate with our hopes and expectations.

Earlier in the chapter I noted that on a modern understanding of electromagnetism—that is, one without an aether—the electromagnetic field is just a new kind of thing that can be arranged in various ways. These different arrangements turn out to correspond to very familiar phenomena, like light, radio waves, or Wi-Fi signals. Whatever else it is, light seems to be a kind of stuff. A world filled with light is not really empty. Similarly for radio waves, X-rays, Wi-Fi signals, and so on. More importantly, the electromagnetic field counts as matter in general relativity. The various configurations of the electromagnetic field have different amounts of energy and momentum associated with them, and that energy and momentum influence the curvature of space-time just as the sun's mass does.

I have now mentioned two surprising examples of empty space-times allowed by general relativity. But there is another class of space-times that are in some ways even more important. These are space-times in which there is no matter at all, in the same sense that troubled Einstein so much when he first encountered de Sitter space-time. Nonetheless, they are filled with waves very much like light or radio waves.

If this is beginning to sound like a riddle, then you're catching on. But let me back up. Very early in the history of relativity theory, Einstein realized that by playing around with his eponymous equation, he could produce new equations that look very much like Maxwell's equations for an electromagnetic field. This was not a big surprise. In fact, Einstein had used Maxwell's theory of electromagnetism as a model for constructing general relativity, so his equation was *supposed* to be similar to Maxwell's.[102]

These similarities, though, had an unlooked-for consequence. We already know that Maxwell's equations describe electromagnetic radiation—that is, configurations of the field that look like waves that travel from one place to another. After all, this is precisely what happens when light from a lightbulb travels to your eye or a signal from your Wi-Fi router travels to your computer. What Einstein realized was that the similarities between his equation and Maxwell's equations meant that general relativity permitted something very similar: waves that could travel from one place to another. We now call these *gravitational waves*.

When Maxwell first introduced electromagnetic radiation, he understood it as a particular sort of oscillation in the aether. Later, with the advent of relativity, these waves were reinterpreted: now, the electromagnetic field was itself a kind of entity, and it could oscillate in various ways to produce familiar electromagnetic phenomenon. But what about gravitational waves? If these are supposed to be oscillations of some sort, then *what is oscillating?* What plays the role of the electromagnetic field?

The answer is that gravitational waves are oscillations in the geometry of space-time itself.

What does this mean? Remember that space-time geometry, as encoded by the space-time metric, determines things like lengths and angles in curved space-time. It also determines the default motion of matter, in just the way that the straight lines of Galilean space-time determine how small bodies move. So to say that a gravitational wave is an oscillation in space-time geometry is to say that when a gravitational wave passes through some region of space-time, then the "straightest" lines in that region will oscillate. Small particles just going about their business, without any kind of forces acting on them, will suddenly begin moving as if they are bouncing back and forth, moving toward one another and then away again. Likewise, if you imagined a spring in space-time as a gravitational wave passed by, the spring would alternatively get stretched and compressed by the wave.

Gravitational waves are possible even in empty universes. This means that they are not themselves a kind of stuff, in the same sense that a black hole is not a kind of stuff. Gravitational waves are *also* possible in universes where there is matter—but in those cases, too, they don't count as being matter themselves. And this, perhaps even more than the black hole in Schwarzschild space-time, strains what it means to say space-time is empty.

Think of the spring I just mentioned. When an ordinary metal spring oscillates, it gradually slows down and comes to a stop, as a result of friction within the metal. When this happens, the spring heats up, releasing energy. This energy may be released in several forms, but at least some of it will be as light: that is, as the spring heats up, it will begin to glow faintly. This means that a gravitational wave can transfer energy, first to the spring and then as light and heat.[103]

Here's another example. There are some crystals, known as piezoelectric crystals, that produce a current when they are deformed. This means you can hook them up to a circuit, connected, say, to a lightbulb, and by making them vibrate, you can power the lightbulb. But what if the crystal is made to vibrate by a gravitational wave? Then when the wave passes by, the lightbulb will turn on, apparently spontaneously, even though there was no matter or energy in the ordinary sense influencing the crystal. In both of these cases, not only does empty space-time make ordinary stuff move, these motions can actually produce electromagnetic radiation, which, as I said above, *is* stuff in the ordinary sense. So gravitational waves can *produce* ordinary matter—at least with the right tools.

In other words, general relativity tells us that with the right sort of springs and a passing gravitational wave, it is possible to make something out of nothing.

Despite his early horror, Einstein came around to the idea that empty space-time could have rich structure fairly quickly: although he persisted in his arguments against de Sitter's solution through 1918, by 1920 he appears to have embraced the idea that the geometry of space-time is only partially determined by the matter in the universe—and partly determined by the geometry of the rest of space-time, in ways that lead to complex phenomena even in the absence of matter.

Indeed, in a lecture delivered in 1920, Einstein came to describe the metric itself as a kind of "aether," which both determined space-time geometry and also had some of the features of a material substance, like the electromagnetic field.[104] This close relationship between the metric and the electromagnetic field, which had played an inspirational role in the development of general relativity, now began to take on a new significance for Einstein. He soon embarked on a quixotic project that would occupy him for much of the rest of his career, attempting to build a "unified field theory" that would combine general relativity and electromagnetism in much the way that Maxwell had unified electricity and magnetism. He never accomplished this goal—and indeed, as we will see in the next chapter, the study of electromagnetic theory in the twentieth century took a decidedly different turn.

But while he ultimately accepted the possibility of empty space-times with complex structure, Einstein's views on gravitational waves turned out to be far more complex—and unstable.[105] Although he was the one who first discovered them, his early papers were littered with errors, and as he corrected those errors he began to uncover some very puzzling features of gravitational waves—the most striking of which was that they do not appear to carry any energy, at least in the ordinary sense.

The puzzles were so severe, in fact, that by 1936 he had convinced himself that gravitational waves did not exist at all. He wrote a paper that year with his research assistant, Nathan Rosen, with the title, "Do Gravitational Waves Exist?" endeavoring to show that his own earlier work on gravitational waves depended on various methods of approximation and estimation that he now believed were inaccurate.[106] More principled calculations, meanwhile, showed that gravitational waves actually led to inconsistencies.

He and Rosen sent the paper off to the *Physical Review*, which by the mid-1930s was the most prestigious physics journal in the United States, and arguably the world.[107] Einstein had published regularly in the *Physical Review* for years, and he expected this new paper to be summarily accepted. But something about it gave the journal's editor pause. Instead of accepting it immediately, he sent it to an outside expert, to evaluate it and make a recommendation about its suitability for publication. The reviewer replied with a report running ten pages long—longer than many scientific papers—that was fiercely critical of both the paper's conclusion and several of the arguments made in its favor. *Physical Review* communicated the report to Einstein and Rosen without disclosing the reviewer's identity, indicating that they could not publish the paper in the form submitted unless Einstein addressed the criticisms.

Today, this process, known as peer review, is the norm: it is unimaginable that a paper could be published in a major academic journal without being reviewed by outside experts, and the quality and prestige of a journal is often directly correlated with the rigor of its peer-review process. But as recently as the middle of the twentieth century, this system was still inchoate, and excellent journals would regularly publish articles without any kind of outside review.

Einstein, by this point, was not merely the most famous scientist in the world: he was one of the most famous *people* in the world. In addition to his accomplishments in physics, including his 1921 Nobel Prize, Einstein was an outspoken critic of both fascism and communism in Europe.[108] In 1933, under threats of death from Nazi sympathizers, he fled the Continent—first for the

United Kingdom, and then the United States, where he settled at the Institute for Advanced Study in Princeton, New Jersey. From there, he was a powerful political force, advocating for other Jewish refugees from Germany and urging Western European leaders to take seriously the rising tide of fascism.

He was not accustomed to critical evaluation, much less by an anonymous "expert" on the theory he had himself invented. It is not clear that Einstein even read the report when he received it. Instead, he penned an angry response to the editors of *Physical Review*, dismissing the report and saying that he and Rosen "had sent our manuscript for *publication*," not for review. He withdrew the paper and never published in *Physical Review* again. Instead, he sent the paper to another journal, without even changing the minor typographical errors noted by the reviewer. That journal promptly accepted it.

But the matter did not end there. Einstein's spleen notwithstanding, the reviewer turned out to be right. Although Einstein did not know it at the time (or, perhaps, ever), the reviewer was Howard Robertson. Robertson truly was an accomplished expert on relativity theory, and Einstein acknowledged him as such. Indeed, by this time they both had permanent positions in Princeton—Robertson at the university, and Einstein at the institute. When Robertson reviewed the paper, however, he was on leave in California. Not until he returned to New Jersey, some months later, did he and Einstein speak in detail about the work.

Robertson found that Einstein had not changed his views at all in light of the anonymous report. Nor had he spent any time understanding the criticisms Robertson had raised. Apparently without ever revealing that he had been the reviewer, Robertson gradually convinced Einstein that the arguments in his paper were flawed, so that finally, Einstein completely revised it before it appeared in print. In the final, published version, he argued for the opposite conclusion: once again, he believed gravitational waves were possible in general relativity.

Even after this debacle, however, Einstein expressed periodic skepticism about whether gravitational waves could ever be produced by real physical systems in our own universe. It is not clear he ever reached a stable view on the matter.

Today, there is broad consensus among physicists that gravitational waves are not only possible but are bouncing all around our own universe. Physicists believe they are produced by events that we know occur, such as when a star explodes as a supernova. Many physicists also believe that gravitational waves were produced in the very early universe, shortly after the big bang, and that these are traveling around the universe today. Gravitational waves are usually taken to be one of the main novel predictions of general relativity—one of the ways in which general relativity is most unlike Newtonian gravitation.

Until recently, this would have been the end of the story: though they had been predicted a century ago, and billions of dollars and decades of effort had been dedicated to finding them, no gravitational waves had ever been observed. But then, early in the morning of September 14, 2015, that changed. For the first time, the two separate facilities of the Laser Interferometer Gravitational-Wave Observatory (LIGO), one in Louisiana and the other in Washington, recorded a tell-tale signal of a gravitational wave passing through the earth. Their analysis revealed that the wave had been produced over a billion light years away, when two black holes crashed into one another, sending a ripple through space and time. This discovery provides powerful evidence for general relativity—including the rich, dynamic structure of empty space-time the theory describes.[109]

CHAPTER 3

The Nothing Nothings

When Paul Dirac made his most significant discovery—an equation that formed the basis for his theory of elementary particles known as electrons—he didn't tell a soul.[110] For a month he stayed silent. No letters, no casual remarks to students or colleagues. Nothing. He kept the work to himself, riding an emotional roller coaster between elation at having found something so exquisite and crippling fear that it would all fall apart when confronted with the base facts of experimental physics. He was so scared, in fact, that after making some initial calculations to convince himself of the plausibility of his discovery, he couldn't bring himself to work out any further details, lest he prove himself wrong. To share it with anyone else would only increase the risk that the world was not as beautiful as he'd allowed himself to believe.

No one would have guessed at this inner turmoil—much less its cause—if they had chanced to meet Dirac on the streets of Cambridge, England, during those weeks. Dirac was famously inscrutable, a man of so few words that some of his closest friends and collaborators did not even know his name: the first person to whom Dirac described his secret theory, Cambridge physicist

Charles Darwin (grandson of the other Charles Darwin), had written to him a few months earlier to ask, "What does P. A. M."—the initials with which Dirac signed his correspondence and papers—"stand for?" They had known one another for six years.[111]

His silence was accompanied by a natural asceticism.[112] His rooms in Cambridge were sparse. His study, where he did most of his work, was furnished only with a desk designed for schoolchildren and a decrepit couch for the odd visitor. There were no decorations. His days were spent in a rigid ritual of work: he would rise promptly each morning and work steadily through the day, pausing only for meals. In the evenings, and on Sundays, when he did not work, he would take long walks, usually alone. In 1929, while Dirac was visiting the University of Wisconsin–Madison, a journalist interviewed him for a column in the local paper.[113] When the journalist asked him if he went to the cinema, Dirac replied, in characteristic brevity, "Yes." "When?" the journalist asked. "In 1920—perhaps also in 1930." (In response to an earlier question, concerning whether Dirac could give an overview of his research in layman's terms, Dirac answered, simply, "No.")

Even on the few occasions when he did speak, it was rare for anyone to understand what he was saying. Dirac had a knack for approaching problems in a unique way—and he was unable, or unwilling, to understand other approaches. One either had to learn to think like Dirac, or else forswear communicating with him. Even so, the benefits of learning Dirac's language were manifest very early in his career. He finished his PhD at Cambridge in 1926, at the age of twenty-three. His was the first dissertation ever written on the new quantum theory, which was just then in the pangs of a difficult and complicated birth. He immediately won a fellowship to travel to Copenhagen, Denmark, and Göttingen, Germany, two of the leading centers for research on the new theory. Although he was not universally beloved—his German colleagues in particular found him difficult, both intellectually and personally—he commanded immediate respect.[114]

He returned to Cambridge in October 1927, having just been elected Fellow of St. John's College, Cambridge. He found the secret equation soon after, sometime in late November or early December of that year. He was just twenty-five years old, but he was already one of the most famous scientists in Europe and arguably the greatest English physicist since Newton.[115] (Maxwell, recall, was a Scot!) By twenty-nine, he would be elected Lucasian Professor of Mathematics at Cambridge University—the same position that Newton was elected to in 1669.

A little over a year after that, at the age of thirty-one, he was awarded the Nobel Prize, for using his equation to successfully predict the existence of an entirely new form of matter. Along the way, he developed a quantum theory of electromagnetism that, much like Maxwell's theory before it, forever changed how we understand the physics of nothing.

A spirit of radicalism and revolution accompanied the new quantum theory in the mid-1920s. Like most revolutionary movements, it was a young man's game. While Dirac was one of the youngest to make significant contributions during the heady days when the theory went through its first major growth spurt, many of the other major players—Werner Heisenberg, Louis de Broglie, Wolfgang Pauli, Pascual Jordan—were also in their twenties or early thirties. Its youthful inventors knew they were doing something extraordinary, and they were eager to make their mark.

Among older physicists, however, there was considerable skepticism—when they paid any attention at all. Paul Ehrenfest, a physicist in Leiden, Holland, who regularly visited the quantum hotspots in Berlin and Göttingen, used to travel with a Ceylonese parrot trained to say (in German), "But gentlemen, this is not physics!" Ehrenfest was fond of suggesting that the parrot should lead discussions of quantum theory.[116] (Even so, he was a major contributor to the theory.)

The parrot's sentiments were understandable. The new quantum theory presented a view of reality so dramatically different from anything that had come before that many physicists, Einstein included, believed it couldn't amount to a complete picture of reality at all.[117]

The theories I have discussed in previous chapters differ in many ways. But they all have one basic feature in common. In all of them, it is natural to think about a given state of affairs—that is, a configuration of physical objects, at an instant or over time—by imagining bodies, with more or less defined boundaries, located somewhere in space (and time). So, for instance, in Newtonian gravitation I can represent space as an infinite three-dimensional "container" of possible locations, and at any instant there are just some ordinary things, like planets and baseballs, lying around in it. If I move to a space-time framework, either classically or in relativity theory, I simply imagine these same sorts of things occupying different positions at different times, snaking through four-dimensional space-time.

This way of characterizing a state of affairs, or, as physicists often shorten it, a *state,* has intuitive appeal. After all, our visual fields are laid out something like this: when we look around the world, we see objects, of more or less definite shapes and sizes, arrayed in space. The way we represent states in space or space-time is exactly analogous, except that we try to adopt a "view from nowhere" perspective, characterizing the spatial (and spatio-temporal) relationships between things independently of what any particular person would see if she opened her eyes.

In quantum theory, this picture of a state as a definite arrangement of objects in space or space-time goes out the window.[118] A quantum state is a much more abstract thing. (Note that I am gradually shifting from the ordinary usage of "state of affairs" to a more technical one, characteristic of quantum theory.) We still think of a state as a possible configuration of physical stuff. But these possible states of affairs no longer correspond in any direct way to a definite arrangement of bodies in space and time. Indeed, one cannot "read off" of a quantum state definite information about, for instance, where your chair is located or how fast a bus is moving, in the way you could from a comparable description in Newtonian or relativistic physics.

So what *does* a quantum state tell us? Rather than straightforwardly characterizing locations and velocities, a quantum state encodes *probabilities* about what you might find if you were to try to measure some physical quantity. The quantities that we might in principle measure—such as the position of an object, its momentum, or its charge—are known as *observables*. So a quantum state is a way of determining how likely a given result would be if we performed a measurement of an observable.

Of course, a *classical*, that is, nonquantum state of the sort described above may also be thought of in this way. Think for a moment about the classic video game *Pong*. *Pong* was one of the first video games ever developed, released in 1972 by the now-defunct software company Atari.[119] The game is inspired by tennis or ping-pong: the screen is divided into two halves, one for each player. There is a small "ball" (actually, just a few pixels) that bounces around a screen. Each player has a "paddle" that they can use to hit the "ball," with the goal of preventing the ball from passing outside the boundary of the screen on his or her own side. With decent players, the ball can bounce around the screen for a long time.

The way the ball's position is represented in *Pong* is exactly how a real tennis ball's position would be represented in classical physics: by specifying its location in space over time. This gives us the information we need to determine the probability that the ball will be located in a given place at a given instant. Now, if we imagine simply recording the position of the ball at each instant of the game, the sorts of probabilities we extract won't be very interesting: at each instant, the ball is simply located somewhere. So the probability of finding the ball in *that* location, according to our classical representation, is 100 percent. Likewise, the probability of finding the ball in any *other* location at that instant is 0 percent. After all, the ball is either *there* or not. And if the ball is there, and we try to observe whether it is there, then we are certain to find it; otherwise, we are certain not to!

That said, there are some situations where, even classically, we might want to think of a state as assigning more interesting probabilities to the ball's position. For instance, the ball begins every round on the line dividing the two halves of the screen. But it does not always begin in the same place along that line, nor moving with the same speed or in the same direction. One might think, then, that we can represent the initial position of the ball by assigning probabilities to various possible initial configurations. In this case, it would be very useful to think of the state of the game at the beginning of the round as something that just assigns probabilities to the ball's having a given location.

We can also ask how those probabilities might change over time. At the very beginning of the round, the probability that the ball will be located near one of the paddles, away from the middle line, is still 0 percent. But a few moments later, the ball will have moved, and the probability of its being located in any particular place now will depend on the initial probabilities concerning its position and velocity.

It is important to be clear about what these probabilities are recording about the world. The reason we might want to think about the probability of the ball being located in a particular place a few moments after the game begins is that we don't *know* where it will begin the game. In other words, the probabilities reflect our uncertainty about the ball's position.[120] But if we simply had more information—for instance, if we knew for sure where the ball would start, and with what initial speed and direction—then we could be certain about where the ball would be a moment later, or even a few moments later, at least until something happened that might introduce new uncertainty, such as a player hitting the ball with a paddle.

Most important, the fact that we might sometimes think of a classical state of affairs using probabilities does not suggest that at any moment of the game the ball doesn't *have* a position; all that's at issue is whether we happen to know what the position is. It's all about us, not the world.

The probabilities that states assign to observables in quantum theories, on the other hand, are a completely different kettle of fish. In quantum theory, we can know everything there is to know about the ball and yet still have the state be such that we have a 50 percent chance of discovering that the ball is on player 1's side of the dividing line and a 50 percent chance of discovering that it is on player 2's side. Of course, these probabilities *do* mean that we are uncertain about where we will find the ball. But the reason may not be that we lack some relevant information. Quantum theory also permits another possibility: that the ball simply isn't located in *either* place—nor, for that matter, anywhere else.

In other words, quantum theory says that a ball, which is the sort of thing that certainly can have a position, and which, whenever we try to look for it, does have a position, may not always have a position.

I said in the previous chapter that relativity theory gave us some weird results. But quantum theory is far, far stranger.

One of the very strangest things is immediately on display in this example. I said above that, according to quantum theory, if you ever try to look for the ball—that is, if you "perform a measurement" of the ball's position—you will find that it does have a position, just as in classical physics. But in between measurements, this need not be true. (Once again, this is not just a matter of uncertainty about where the ball has gone!) The big mystery here is this: What makes a measurement so special? And what counts as a measurement? Can only people measure things? Can dogs? Robots?

Shockingly, although quantum theory in its modern form is nearly a century old, no one knows the answer to these questions. This basic problem, which comes up in a number of different ways throughout the theory, is known as the *measurement problem*. Lots of possible solutions have been proposed, ranging from changing our theory of probability, to creating alternative theories that make similar predictions to quantum theory, to imagining that there are many parallel worlds, in which everything that *can* happen *does* happen.[121]

All of these have their defenders, but there is no consensus among them—and all of the known approaches face serious problems of their own. But until we have a satisfactory solution to the measurement problem, it is fair to say that no one really understands quantum theory.

Dirac first arrived in Cambridge as a graduate student in 1923, less than a decade after Einstein had published his first papers on general relativity. Even special relativity was still a fresh topic. Dirac was drawn to relativity theory from the beginning of his career. One of his early mentors in graduate school was Arthur Eddington, the most influential advocate for Einstein's new theory in the United Kingdom. Eddington worked tirelessly to translate Einstein's ideas into English—and also into terms that physicists, few of whom had the mathematical training to appreciate Einstein's work, could understand. In 1919, Eddington led an expedition to the island of Principe, off the west coast of Africa, to photograph a solar eclipse. These photographs were crucial in establishing Einstein's theory because they showed that the gravitational influence of the sun curved the light from distant stars in the way predicted by Einstein's theory—and not by Newton's.[122]

Although Dirac's attention turned quickly to the new quantum theory, he maintained an interest in relativity, and much later in life, he published an influential textbook on the subject. Even within the quantum domain, many of his most important contributions concerned relativity theory. When quantum theory was first developed, it was customary to take for granted a Newtonian conception of space and time, since this simplified many calculations. Dirac, however, wanted to develop a quantum theory of the electromagnetic field—a project that forced him to reckon with the geometry of special relativity, since, as we saw in the previous chapter, electromagnetism and relativity theory are intimately linked.[123]

In early 1927—about nine months before he found his theory of the electron—he published his first paper combining quantum theory and electromagnetism. In it, he coined the name "quantum electrodynamics"—QED for short—for his new theory.[124] It was the first major attempt to develop a quantum theory of *fields*, the entities Maxwell had introduced into physics in the nineteenth century, and which played a crucial role in Einstein's thought. And it was in this paper that Dirac took the first steps toward a completely new understanding of the physics of nothing—though a full understanding of empty space in relativistic quantum theory was still decades away and required many more pieces.

As Dirac began work on QED, Einstein was still very much a presence in physics. More, he was already a celebrity, both in physics and the world at large. He had won the Nobel Prize in 1922, shortly before Dirac started graduate school. But he was not so old and famous that he wasn't paying close attention to the work Dirac and his cohort were doing. Like Ehrenfest, Einstein was a member of the older generation of physicists who made substantial contributions to quantum theory and followed the new theory with interest—but even more than Ehrenfest, Einstein held the developments of the 1920s at arm's length.[125]

Dirac and Einstein met in 1927, at the fifth Solvay conference, in Brussels, Belgium.[126] Ernest Solvay was a Belgian chemist and industrialist who began funding major international conferences in 1911. The first Solvay conference, in 1911, was one of the first occasions for a sustained discussion of early quantum mechanical ideas. Einstein, whose paper on light quanta in 1905 had jump-started these early efforts, was the second-youngest physicist there. He was a young radical, defending both the new theory of relativity and the methods of early quantum theory among physicists who had made their careers in a thoroughly classical world.

By 1927, however, Einstein had become the old guard. At this conference, Dirac was the youngest participant, and Einstein the curmudgeon. Given Dirac's intense shyness and Einstein's stature, they did not engage much directly. Instead, Dirac looked on as Einstein argued with the other physicists present—most prominently, Niels Bohr, a member of Einstein's own generation who was much more enthusiastic about the turn quantum physics had taken.[127] Bohr had positioned himself as a kind of godfather to the new quantum movement, and many of the more junior physicists looked to him as their leader. (It was Bohr whom Dirac had visited in Copenhagen in 1926 after finishing his PhD.)

Later, Dirac reported that although he listened to these debates, they bored him.[128] Both Einstein and Bohr were deeply interested in philosophy: they each wanted to reconcile the new physics with his own conception of what the world is like and how we come to know it. They had fundamental philosophical disagreements that influenced their views on the physics.

Dirac, meanwhile, wanted to separate himself and his work from philosophical issues, which he believed were distractions. As far as he was concerned, developing the right mathematics was the primary goal, and interpreting that mathematics in a philosophically satisfying way was secondary, if one ever got around to it at all. Neither Einstein nor Bohr was saying anything that could be translated into precise mathematical claims. It was all just *talk*, and Dirac was never one for talking.

Einstein got under Dirac's skin in another way, too. Einstein had a habit of expressing himself with references to God. (His worries about the probabilistic aspects of quantum theory are often summarized with one of his most famous quotes: "God does not play dice."[129]) For Dirac, littering a conference on physics with claims about what sorts of things God would and wouldn't license just underscored his more general concern about how philosophical and interpretational matters were interfering with the real work.

According to Heisenberg, an attendee of Dirac's own generation, it was these references to God that ultimately provoked an outburst from Dirac. One evening, during a conversation in the hotel lounge between the younger participants at the conference, the ordinarily silent Dirac surprised everyone by announcing, "Religion is a jumble of false assertions, with no basis in reality. The very idea of God is a product of the human imagination."[130]

In response, Wolfgang Pauli, another member of the younger set, quipped, "There is no God, and Dirac is his prophet." Even Dirac laughed.

Attributing any major scientific idea—let alone whole theories—to a lone genius is always an oversimplification. But how much of an oversimplification is a matter of degree. Of course, to hear Leibniz or Hooke speak of it, Newton merely cribbed ideas from them. And without the well-timed interventions of Minkowski and Grossman, Einstein might never have found his way to general relativity.[131] Even so, while Einstein and Newton certainly drew on the work of other scientists, their own contributions to the theories of space, time, and gravitation that bear their names far outpaced those of even their ablest contemporaries. It is not so misleading to focus on their work.

Quantum theory is a different matter. It would be a huge distortion to attach a single name to it. The first inklings appeared in papers by German physicist Max Planck published 1900, with major early contributions by Einstein, Niels Bohr, and the German physicist Arnold Sommerfeld.[132] But none of this work presents a definitive statement of quantum theory remotely analogous to what one finds for Newtonian gravitation in the *Principia* or for relativity in Einstein's papers of 1905 and 1915. Instead, these early works in quantum physics contain startling ideas applied to specific problems, but without the grand architectonics of a new theory.

It was not until 1925 that something like a full theory began to emerge, in a series of very rapid developments. It was produced piecemeal, and several people stand out as major contributors—Werner Heisenberg, Erwin Schrödinger, Max Born, Pascual Jordan, Wolfgang Pauli, Paul Ehrenfest, and of course, Dirac. Ultimately it was Dirac and Hungarian polymath John von Neumann who showed how all of the pieces fit together, by synthesizing the new theory in influential textbooks published in 1930 and 1932, respectively. These books, the first serious attempts to systematically present a quantum theory deserving of the name, were the starting point for essentially all subsequent work on the subject. But in both cases, the principal task was to pull together three decades of research into a coherent whole, and not to invent a new theory from scratch.

QED was originally just one strand in this tangle of new ideas. Although Dirac coined the term in his 1927 paper and was the first to take a stab at developing a systematic theory, he was really just picking up where earlier researchers had left off. In fact, the very first ideas about quantum theory, from Planck and Einstein, were already ideas about QED: Einstein's 1905 paper, for instance, was all about the quantum properties of electromagnetic radiation. It was there that he introduced the idea of "light quanta," which would become a core concept in Dirac's theory.[133] So when Dirac began to work on QED in the fall and winter of 1926–1927, his goal was at least in part to bring this older work on electromagnetism in line with the newest developments in quantum theory.

Dirac made remarkable progress, even in that first paper on QED. But he didn't come close to finishing the project. It was another physicist, Pascual Jordan, who took the next major step in inventing the theory.[134]

Jordan was based in Göttingen, along with Heisenberg, Born, and Pauli, with whom he actively collaborated. He was the only person to make major contributions to quantum theory during the heady period in the mid-1920s who was younger than Dirac—which he was by just two months. (Even so, he finished his PhD in 1924, two years before Dirac did!) Like Dirac, he was retiring and difficult to understand—though in Jordan's case, the problem was a severe stutter, not a refusal to speak. The two men met in 1927, when Dirac traveled to Göttingen from Copenhagen. (They effectively switched places for most of the visit, however, as Jordan went to Copenhagen that spring). This meeting occurred during the very height of early activity on QED, and there was a sense of intense competition between the awkward young men.

Jordan had already done some preliminary calculations in 1926, in a famous paper with Born and Heisenberg that laid the foundation for Dirac's QED paper.[135] But his most important contribution came in the months after Dirac's paper appeared. It concerned the wavelike nature of matter.

One of the core ideas of quantum theory is that ordinary matter sometimes behaves as if it were made out of tiny bits of discrete stuff—particles—and sometimes behaves as if it were some kind of wave, like the electromagnetic field. This puzzling situation—the response to an apparent either/or question turned out to be "both" and also "neither"—led to some mystical discussion of "wave-particle dualism" in the quantum domain. It also led to some concrete theorizing concerning the wavelike properties of otherwise particulate matter. In 1925, following up on earlier ideas due to Louis de Broglie, Erwin Schrödinger proposed an equation that seemed to capture these wavelike properties of stuff.[136]

Schrödinger's equation was originally proposed as part of a version of quantum theory that he called "wave mechanics." The central posit was that matter actually *is* a wave, just like electromagnetic radiation. This interpretation was quickly challenged, however, and soon a new, probabilistic interpretation of Schrödinger's waves was proposed by Max Born.[137] Today, following Born, physicists believe that Schrödinger's waves do not describe matter directly, but represent states of affairs, in the sense described earlier. The wavelike properties of matter arise because the probabilities associated with their states oscillate.

Jordan had another idea about how to further develop Schrödinger's theory—one that ended up going in a quite different direction. Jordan's idea was that ordinary matter, including electrons, should be treated as wavelike for the purposes of quantum theory. This may sound like Schrödinger's idea all over again, but in fact it went further. To see how, one needs to understand that a common procedure for developing quantum theories, both in the 1920s and today, is to begin with a classical theory—say, Newtonian particle mechanics or Maxwell's electromagnetism—and use that as a model for constructing a corresponding quantum theory. (This is what Dirac did to arrive at QED.) From this perspective, Schrödinger's wave mechanics was supposed to be a *quantum* theory—specifically, one corresponding to a classical theory of a particle.

Jordan's idea was that to build a quantum theory of electrons, he should begin with something like a Schrödinger wave at the *classical* level. He could then proceed to construct a quantum theory of electrons using strategies similar to the ones Dirac used for QED. In other words, first Dirac showed how one could develop a quantum theory of electromagnetism—which, even classically, was conceived as a field. Then Jordan argued that other stuff, too, including matter that classically seems to be particlelike, should be treated as fields, just like the electromagnetic field.[138] Developing a quantum theory of electrons, then, would amount to developing a quantum theory of an appropriate electron field along the lines of what Dirac did for the electromagnetic field.

But what should that "electron field" be like? One option would be to look to Schrödinger's waves. But in fact, this would not work, because Schrödinger's equation was incompatible with relativity theory. Much like Newton's law of gravitation, Schrödinger's theory appeared to allow *instantaneous* action at a distance. In other words, it presupposed some special standard of simultaneity, of precisely the sort that relativity theory ruled out.

It was because of these difficulties with the Schrödinger equation that Dirac attempted to develop a new wave equation, analogous to Schrödinger's equation, but compatible with relativity.[139] In fact, even as he set out to do this, others believed it had already been done: a year earlier, in 1926, several physicists proposed a relativistic analogue of the Schrödinger equation, now known as the Klein-Gordon equation.[140] (It was clear even before Jordan's proposal that *something* would need to be done about the incompatibility between the Schrödinger equation and relativity theory. Indeed, Dirac himself was primarily motivated by this general incompatibility, and not Jordan's arguments in particular.) When Dirac happened to mention to Bohr at the Solvay conference in October 1927, just a month before his discovery, that he was looking for a relativistic equation to describe the electron, Bohr announced that this problem had already been solved and advised him to work on other things!

Dirac was unmoved by Bohr's advice. He had technical reasons for thinking the Klein-Gordon equation was inadequate for a theory of the electron.[141] And it turned out that Dirac was right:

today, we would say that the Klein-Gordon equation describes a very different kind of particle, with measurably different properties from electrons. But it would have been almost impossible to articulate what the problems with the Klein-Gordon equation were before Dirac's own equation had been introduced, because it wasn't clear what a satisfactory equation for an electron would even look like (not that articulating such things was ever a priority for Dirac). It was a feat of remarkable physical intuition for Dirac to see both the inadequacies of the Klein-Gordon equation and a path forward.

Dirac's secret theory of the electron didn't stay secret for long. By Christmas 1927, about a month after he discovered it, he mentioned it to Darwin, who promptly wrote a letter to Bohr detailing the discovery. On New Year's Day 1928, Dirac submitted an article describing what he had done to the *Proceedings of the Royal Society*.[142] Even then, he kept a tight lid on it, barely mentioning it even in letters to his friends and former collaborators in Göttingen.

Still, word spread quickly that Dirac had accomplished *something* great—even if no one knew quite what—and the secrecy only fueled anticipation. By the time the paper appeared in print, in early February, everyone in quantum-dom was anxious to read it.

When Dirac's equation finally appeared, it was immediately recognized as a major accomplishment. (Jordan, who had been trying to develop his own theory of the electron, fell into a deep depression and essentially left physics.) Although Dirac was too scared to perform the calculations himself, others quickly established, in some detail, that the theory of electrons based on his equation could correctly reproduce several important experimental results.[143] So although it was a startlingly original mathematical development, it was immediately brought into contact with experiments. Things looked very good.

But this didn't mean there weren't problems. As Dirac himself quickly noticed—and as others pointed out to him—even though his equation seemed to describe electrons correctly, it described *too much*. It also described another kind of particle—one that no one had ever seen.[144]

This interpretation of the equation, as describing two kinds of particles at once, wasn't seen right away. Originally, the equation was understood to say that there are possible states for electrons that are very different from the states electrons were usually observed in. These appeared to be states in which an electron does not have positive energy—and indeed, in which the energy could become arbitrarily negative.

These negative energy states led to serious problems, both experimentally and conceptually. For one, electrons in such states would behave very strangely according to the theory, such an electron would behave as if it didn't have the electric charge of an electron at all. Instead, it would act as if it had the *opposite* charge. In all other respects, however, it would have been identical to an electron. No electron had ever been observed to behave this way in an experiment. And since objects tend to gradually lose energy, moving to ever-lower-energy states, one might have imagined that if these negative energy states were really possible, we would see them all the time.

It was not obvious what to make of this situation. Dirac's own opinion shifted several times over the next two or three years. But by 1931, he had come to a settled view. These negative energy states, he declared, corresponded to a "new kind of particle, unknown to experimental physics, having the same mass and opposite charge to an electron."[145] He called these new particles "anti-electrons," though others quickly dubbed them "positive electrons" or "positrons." They were the first example of something that we now call "antimatter."

Antimatter is a fancy name that can conjure up all sorts of things. (And indeed, Dirac, at least initially, imagined antimatter as a kind of matter-shaped "hole" in a sea of ordinary, albeit negative-energy, electrons.) But for all intents and purposes, positrons are just a new kind of matter, identical in all ways to electrons, except that a positron reacts to the electromagnetic field in exactly the opposite way as an electron. If an electron would go up, a positron would go down, like a contrary twin.

When Dirac made this suggestion, it was radical. No one had ever seen such an entity in any experiment. It seemed as if Dirac was making up some new kind of stuff just to keep his beautiful equation alive for a little longer. Positing new kinds of particles was a cheap way out, a refusal to play by the rules of the game, which were to develop theories of the phenomena that experimentalists actually observed.

But physicists' views on the positron changed very quickly, for a simple reason: they started showing up in photographs.[146] In August 1932, Carl Anderson, a postdoctoral researcher at the California Institute of Technology, in Pasadena, California, was studying cosmic rays, which are very-high-energy particles that occasionally bombard the earth from outer space. He was using a device known as a cloud chamber, which is essentially a closed box filled with water vapor. When a charged particle passes through a cloud chamber, it produces a mist along the path it travels, leaving a visible trail that can be photographed and studied. By looking at the properties of these paths, one can deduce what sorts of charged particles had passed through the chamber.

Anderson quickly noticed that some of the tracks in his

chamber were behaving strangely. The tracks looked exactly like the tracks of electrons. But they were going in the wrong direction in a magnetic field—just as one would have expected if they were positrons. Within a year, other physicists managed to produce similar images in their cloud chambers.[147] By early 1933, skepticism had been replaced by awe: Dirac's radical proposal had become an astonishingly accurate prediction. The Nobel Prize followed that November.

The discovery of the positron notwithstanding, the breakneck pace of progress in the late 1920s slowed by the early '30s, hampered both by difficult problems in the new theory and by the tectonics of global politics. First came the Depression and the sorrows of Weimar Germany, and then the rise of National Socialism in Germany and the beginning of World War II.[148] Like Einstein, many scientists were displaced during the 1930s, fleeing the Nazis; then they were conscripted to work on military projects during the war, with many erstwhile collaborators pitted against one another on opposite sides of the hostilities. And so, as it happened, the leaders of the quantum revolution in the 1920s were not destined to complete the project they had begun. It was a younger generation of physicists, working in the period just after the war, who would carry QED to the next waypoint.

Leading the way were two Americans: Richard Feynman and Julian Schwinger.[149]

Dirac's positrons are striking because of their contradictions. We call them *anti*particles to capture the idea that they are the opposite of particles. But in fact, they are opposites only in charge. In other regards they are identical to their particle twins: a positron has precisely the same mass as an electron, precisely the same statistical properties, and it is influenced by just the same forces. Indeed, it is these similarities that make the opposite charges so interesting. If particles and antiparticles were not so very close to being the same, the fact that they are different in such an important way would not be so remarkable.

The same could be said of Newton and Leibniz. Nearly exactly the same age but born on opposite sides of the Continent; independent discoverers of the same revolutionary mathematical theory, but with strikingly different formalisms and emphases; both polymaths drawn to similar problems in mathematics, physics, and theology—and yet each profoundly critical of the other's work. They demand to be considered together, to be compared. If they weren't so startlingly similar, their disputes would hold far less interest.

For Dirac, there was Jordan: again, nearly exactly the same age—and startlingly young when they made their most important contributions, even in a field of young geniuses. Admiring rivals, they worked on nearly the same problems, and in at least one case independently came to the same solution. Between the two of them, they invented the quantum theory of fields.

Feynman and Schwinger were another of these pairs. Both were native New Yorkers, born within months of one another in 1918 to Jewish émigrés from Eastern Europe. Both were prodigies whose remarkable gifts in physics and mathematics were identified when they were still teenagers and cultivated by some of the finest teachers and researchers in the world. Both cut their teeth during World War II, Feynman as one of the chief theorists on the Manhattan Project at Los Alamos, and Schwinger as a radar expert in the Radiation Laboratory at MIT. And then, before and (especially) after the war, both made crucial contributions to the same theory—QED—for which they would share a Nobel Prize in 1965 (with Japanese physicist Sin-Itiro Tomonaga).

These similarities almost certainly contributed to the idea—prevalent among physicists during the 1940s, and apparently accepted, at times, by the two men themselves—that Feynman and Schwinger were locked in an epic race to finish QED.[150] The fact that credit was ultimately shared between them—at least as far as the Nobel committee was concerned—may seem reminiscent of the priority dispute between Newton and Leibniz, now jointly credited with discovering calculus, each of whom felt his own contributions could not be properly appreciated until the other's were minimized.

But in fact the situation was quite different. Despite a sense of competition between these two giants, further progress on QED came in fits and starts, with substantial contributions by Feynman and Schwinger both—as well as by dozens of others, including Dirac, Tomonaga, and Freeman Dyson. More importantly, neither Feynman nor Schwinger ever claimed QED as his own in anything like the way Newton and Leibniz both claimed calculus. They each learned from the other and recognized that they built on one another's insights.

The similarities between Feynman and Schwinger cast in relief some striking differences. One was personality: Feynman was gregarious and extroverted; Schwinger, retiring and shy. After they were awarded the Nobel Prize, Feynman emerged as a celebrity, writing a series of books geared toward a general audience. The first of these were serious attempts to explain modern physics, including QED, to nonphysicists, but by the end of his career, they were collections of anecdotes and amusing stories about his life. (Murray Gell-Mann, another Nobel laureate and one of

Feynman's colleagues at CalTech, once referred to the first of these snidely as "Dick's joke book."[151])

Schwinger, meanwhile, receded from the limelight after winning his Nobel Prize, devoting much of his career to developing new, and increasingly idiosyncratic, lines of research. While Feynman grew into a living legend, Schwinger became increasingly isolated. In 1971, he left Harvard, where he'd been since the end of World War II, for UCLA, demoralized by colleagues who criticized his later work as too formal and obscure. By the late 1980s, he was deeply involved in research on cold fusion that lay so far from the mainstream that he could not publish it in major journals.[152] He ultimately resigned from the American Physical Society—the major U.S. professional society for physicists, and publisher of the *Physical Review*—in protest, insisting that they were stifling academic freedom and suppressing his work. It is a significant indication of where his career had ended up that this act of defiance attracted little interest from the physics community.

These differences in personality were also manifest in the men's early work. One of the ways in which Feynman and Schwinger were similar was that, like Dirac, each insisted on developing his own ways of thinking about problems—and ultimately, his own personal language for understanding the world. As a result, they approached QED from completely different directions—as Feynman would put it, they "climbed the same mountain from different sides"—and during the height of activity on QED, they often couldn't understand one another.[153] They could, however, compare results. That they consistently got the same answers, they thought, meant both must be onto something real and important.

For instance, Feynman had a distinctively geometrical way of solving problems. One of his most important contributions to

QED was a method of calculation now known as Feynman diagrams, which allowed physicists to quickly compute solutions by imagining particle trajectories through space and time. Schwinger, meanwhile, preferred to think formally and algebraically, performing calculations by manipulating symbols on a page rather than drawing pictures. (In this way, Schwinger was far more like Einstein and Feynman like Dirac—which is surprising, given that Einstein's theory may be the most important "geometrical" theory in physics and Dirac managed to write a textbook on relativity theory without any diagrams![154]) Schwinger agreed that Feynman diagrams could be helpful, but he felt that they obscured both the math and the physics. He did not teach them in his courses at Harvard, and students were strongly discouraged from using them—though they did so anyway.[155]

Schwinger and Feynman also had different ideas about what the basic building blocks of nature really are. In a sense, they represented opposite sides of a dispute that goes to the core of quantum field theory—and one that is essential to understanding empty space in QED.

Feynman's first step on the road to QED was to develop a theory of electromagnetic interactions in which the electromagnetic field played no role at all: instead, what one had were particles—tiny, pointlike bits of matter with no internal structure—that acted on one another at a distance.[156]

It might seem surprising that such a theory is even possible, given that general relativity was developed in part to avoid the fact that in Newtonian gravitational theory, gravitational forces acted at a distance. But the real problem with Newtonian gravitation was that gravitation acted *instantaneously* at a distance, which no longer made sense in the context of special relativity. Feynman's particle-only theory didn't have this feature: instead, particles acted on one another with a delay that just happened to be the time it would take for electromagnetic radiation—or a photon—to propagate from one particle to the other.

In order to get the theory to work, he needed to add a strange-looking piece: he needed to allow for particles to act on one another both forward and *backward* in time, so that the motion of a particle right here and now depends not only on the distribution of other particles in the past, but also on the distribution of other particles in the future. It was this particles-only theory that Feynman tried to turn into a quantum theory of electromagnetic interactions, and which ultimately led him to solve problems in QED.

Schwinger, on the other hand, wanted to start with fields as the basic entities: a quantum theory of the sort of entity that Maxwell had introduced into physics, and one that took seriously the picture of matter that Jordan had proposed.[157] The theory he produced seemed completely different from Feynman's: instead of particles interacting at a distance, there was a new kind of entity called a *quantum field*. This was meant to be like an electromagnetic field, but one whose possible states were understood in the quantum way described above—that is, as assignments of probabilities to observables. This quantum field could have various kinds of "excitations," which were to be thought of as analogous to different oscillations in an electromagnetic field.

Particles were not a native part of the theory at all; instead, particlelike phenomena were to be thought of as special sorts of excitations in this quantum field.

To someone looking at these theories in the late 1940s—including Feynman and Schwinger themselves—they seemed as different as could be. They were theories of different sorts of stuff in the world—particles, and not fields, in one case; fields and not particles in the other—according to which different sorts of things could happen. They gave different explanations of phenomena, seemed to face different problems, and were motivated by very different intuitions about what the world must be like. And yet they made the same predictions for what one would see in laboratories. In fact, they produced more or less identical calculations for all sorts of physical quantities, observable or not.

Today, physicists have reached a kind of détente between the particle and field views.[158] They tend to talk like Feynman. We build *particle* detectors and *particle* accelerators, to study the properties of electrons, quarks, and Higgs bosons—all examples of elementary *particles*, according to our current best theories. And yet it is actually Schwinger's picture of material reality that forms the core of modern quantum field theory. The particles of modern particle physics are not particles in the classical sense of small, localized bodies bouncing around like tiny ball bearings. They are manifestations of the particular ways in which a quantum field can be configured—much as light is a manifestation of some of the ways in which a (classical) electromagnetic field can be configured, or a tsunami is a way in which the ocean can be configured.

It is a remarkable fact about quantum fields that they exhibit particlelike behavior, and that we can so fruitfully think about elementary particles in a world of quantum fields. Indeed, I continue to speak of particles in what follows. But it is important to remember that "particle" here really describes a particular kind of configuration of a quantum field. The reason it is important is that the idea of a world of particles strongly suggests that there are stretches of empty space between particles. And as we will see, quantum field theory allows no such thing.

In a classical theory, where the states of affairs consist in definite objects occupying definite positions and moving with definite velocities in space or space-time, it is clear what it would mean to say that space-time does not have any stuff in it. (Of course, we also saw in the last chapter that in general relativity, even universes that are empty in this sense can be so richly structured that it's hard to say they are *empty*, full stop. But we will leave that aside for now.) In quantum physics, however, things are less clear. The states are more abstract representations of reality, and the route from a state to physical quantities or measurement is inherently probabilistic. So what would it mean to say that there are no bodies at all?[159]

The answer, at least in quantum field theory, is that the *number* of particles in some region of space and time (or the universe as a whole) should itself be considered a physical quantity—and indeed, one that should be an "observable" in both the technical sense and the ordinary sense of the sort of thing we could, in principle, measure by building a particle detector and counting up clicks.[160] So, given a state, the theory allows us to ask, "How many particles are there?" by asking what the probability is that we would find one particle, or two particles, or three, and so on, if we set up an appropriate experiment. In this way, we can calculate the "average" number of particles that we should expect to find if we performed many similar experiments.

Just as with the other observables described above, however, the probabilities we get from this particle-number observable are generally not (just) a reflection of our ignorance. In other words, the theory does not yield probabilities merely because we do not know how many particles there are—even though there are, in any given region of the universe, or in the universe as a whole, some fixed number of particles. Not at all. Instead, generic states in quantum field theory are ones in which there simply *is no fact of the matter* about whether there are any particles in a given region at a given time, or even in the whole universe for all of eternity. It is only once we make a "measurement" that we get some sort of determinate number of particles in a region. (Though, of course, the measurement problem still rears up.)

What I just said is true for *generic* states: that is, except for very special cases, quantum states are going to be ones in which there is some probability of finding any number of particles. But there are exceptions: one can find states that are such that it is *certain* that there is exactly one particle in the universe, or exactly two, and so on. There are even states in which it is absolutely certain that there are *no* particles in the whole universe. These special states are known as *vacuum states*.

Vacuum states are the best candidate we have for quantum field theoretic descriptions of situations in which there is no stuff in the universe. They are also essential to the conceptual development of the theory: in quantum field theory, the *other* states one encounters are very often constructed by starting with the vacuum state, and then defining new states that are, intuitively speaking, the vacuum "plus" other stuff. Understanding the vacuum is thus crucial for understanding quantum field theory, including QED.

It is therefore quite striking that, although vacuum states represent situations in which it is certain that if one somehow made a measurement of the number of particles in the entire universe, there would be no particles measured, none of the basic inferences one might make from this situation hold. Most importantly, it simply does not follow from the fact that if we measured the entire universe, we would find no particles, that if we were to go into our own backyards and build a particle detector, we wouldn't find a particle there! In fact, it is perfectly consistent with the universe actually being in a vacuum state that we could go into our backyard and find our gardens and trees and playthings. The whole world as we know it is consistent with the universe being in a vacuum state.[161]

This all sounds very paradoxical. And in a sense it is. After all, if there is nothing in the *whole* universe, how is it that there might be something in *part* of the universe?

What is really going on is that our classical concept of *nothing*, even though it motivated our definition of a vacuum state, doesn't line up with anything that quantum theory gives us. In a sense, we already saw this in the last chapter when discussing the electromagnetic field. Recall that I raised the question of whether an electromagnetic field that had no oscillations anywhere should count as nothing. Is the zero field still a thing? What if I said that the zero field was exactly the same as two waves, moving in opposite directions, whose peaks and troughs exactly canceled one another out? Is that two things or no things?

Perhaps the best way of understanding the counterintuitive features of the quantum vacuum is that quantum theory tells us that even the zero field is very much a kind of *stuff*. And a vacuum state of, say, an electromagnetic field might be thought of as the quantum mechanical analog of the zero field. In other words, in quantum field theory, the vacuum state is just another state *of* matter—one that happens to make some suggestive probability assignments about certain in-principle global measurements, but which can yield quite different results when measured locally by our particle detectors. If we recall that particles are just a particular kind of phenomenon that fields can exhibit in quantum field theory, the vacuum state just represents the field configuration where we minimize the particlelike phenomena. But there is *no* configuration where we can force them to go away altogether.

Physicists will often describe the nonzero probability of finding "something" when we make a measurement in the vacuum state in some small region of space and time as reflecting the possibility that we will catch the field "fluctuating" away from the zero state. A particle detected in the vacuum, then, is a "vacuum fluctuation." The intuitive picture is that the vacuum is a roiling sea of activity—or better, of *possibility*, since the fluctuations concern what could happen on measurement, and not actual events in the classical sense. You might think of static on an old TV or on a radio between stations: there is no signal to speak of—no persistent particles—and yet there isn't silence, either. There's just white noise: random, incoherent background. Fluctuations. In fact, in some experimental scenarios in physics, vacuum fluctuations look *exactly* like white noise.

Dirac's antimatter provides another twist to vacuum fluctuations. One can think of the positive and negative energy states for the electron that Dirac discovered—that is, the states we associate with electrons and positrons—as analogous to the hypothetical electromagnetic waves I described above, whose troughs and peaks exactly cancel one another. This suggests that a pair consisting of an electron and a positron is "equivalent" to a situation with no particles at all: the matter and antimatter effectively cancel one another out in the overall accounting of the amount of stuff in the universe.

In a classical theory, this would not work out, since electrons and positrons both have mass and energy. But vacuum fluctuations provide the little boost of energy necessary for an electron/positron pair to spontaneously pop into existence, travel along for a little while, and then meet again and wipe one another out. The particles created in this way are sometimes called "virtual" particles because, like vacuum fluctuations themselves, they have a kind of indefinite, ethereal nature: they come and go in an instant, little blips in the background that are never really there. But they can nonetheless influence other physical processes.

Thinking of the vacuum in terms of fluctuations and spontaneous electron/positron creation and annihilation, as if these were events that randomly occurred around the universe, is just a heuristic. The theory tells us about the probabilities of finding particles if we try to detect them, but it does not tell us anything definite about what those particles are doing, or even whether they are definitely *there*, in between measurements. Even so, it's a useful picture to have in mind when trying to understand what nothingness is like in quantum field theory. In a classical theory, it would be the blackness of a turned-off TV; in quantum field theory, it's the static of no channel.

The features of the vacuum state in quantum field theory that I described above—a chance of finding stuff even when there's nothing there; matter and antimatter popping into existence only to disappear again a moment later—are so perplexing as to verge on unbelievable. Surely, you might think, the theory can't be right. It sounds like speculation and mysticism. Worse, it isn't even coherent.

But in fact, the peculiar features of the QED vacuum are experimentally testable—or at least, they have testable consequences, of a sort that are now commonplace in experimental physics. One important example is known as *vacuum polarization*.[162] Vacuum polarization was first noticed in 1934, by Dirac and others, as they began to understand how complex the structure of the vacuum in QED really is. But physicists working in the '30s were unable to extract predictions from the theory as they understood it then.

Instead, the first theoretical calculations correctly describing an observable consequence of vacuum polarization—an effect known as the Lamb shift—were done by Hans Bethe, Feynman's old boss in the theory division at Los Alamos during World War II. The story—which has taken on the status of legend in physics—is that Bethe did the calculations on the train, on the way home from a conference on Shelter Island, New York, in 1947.[163]

This conference was a major turning point in the development of QED. It was the first large conference after the end of the war. Ideas that had been developed as part of the war effort—most of which were treated as military secrets—could now be discussed among civilians. Both Schwinger and Feynman were there, and it was the first time that both were able to present their newest ideas on QED. Bethe reported being so excited by the talks he heard, particularly from this younger generation of physicists, that he couldn't stop thinking about the new ideas.

Most of what I have said so far about light and electromagnetic radiation has been about what happens when there is nothing else around: that is, light traveling in (otherwise) empty space. But electromagnetic radiation can also pass through some materials, like glass and water.[164] When this happens, the behavior of the electromagnetic field within the medium changes. This is because ordinary objects are made up of atoms, which in turn are made up of protons and electrons. These carry electromagnetic charge, which means that they both produce, and are themselves affected by, excitations in the electromagnetic field. So when light passes through something like water, it interacts with these particles, and that changes how the light travels. The reaction of the medium to the radiation passing through it is known as the *polarization* of the medium. This sort of interaction explains lots of ordinary phenomena, from the partial reflectiveness of windows to rainbows.

The basic idea behind vacuum polarization is this: if the vacuum in QED is rife with virtual electron/positron pairs, as the developing theory of QED indicated, then light should interact with these particles, just as it does with the electrons and protons in glass or water. And this interaction should affect the electromagnetic field in (otherwise) empty space.

Now, polarization is usually measured by studying the properties of radiation passing through a medium. But the effects are described relative to the motion of light in empty space. Thus, if empty space is itself polarized, there is no direct way to tell by studying how light moves, since there is no further standard against which to compare the motion of light. But Bethe realized that there are more subtle ways of testing the effect. In particular, the same mechanism should also affect how the electromagnetic field behaves near isolated charged particles, as a correction to the classical relationship between a charged particle and the electromagnetic field nearby. And *this* effect should be measurable—for instance, by comparing electrons in different orbits around a proton.

Bethe's calculation wasn't exactly a prediction. In fact, two experimental physicists from Columbia University, Willis Lamb and his student Robert Retherford, had attended the Shelter Island conference and described a strange experiment they had recently performed, in which they found a surprising effect in hydrogen. This experiment inspired Bethe to do the calculations he did. But the agreement he found between his calculations and the effect Lamb and Retherford had observed—and which no one at the conference had been able to explain—was close enough that the calculation was taken as a major early success of the theory.

More importantly, it was also convincing evidence that the strange new physics of nothing in QED was telling us something about empty space in the real world—something no one had expected.

QED has been remarkably successful: it is the most accurately predictive theory ever developed in physics, with some predictions verified to ten significant digits or more.[165] And yet there has always been something funny about it—and about quantum field theory more generally. Even as Feynman and Schwinger were busy producing remarkably accurate predictions, members of yet a younger generation of physicists—most notably, Arthur Wightman, Rudolph Haag, and Daniel Kastler—were looking askance at the theory.[166] Sure, they thought, it makes the right predictions—and so, whatever else is the case, it must be getting *something* right. But Wightman, Haag, and others wanted more than that. They wanted to know that the theory could be put on firm mathematical grounds, to establish that it really was consistent and that the inferences other physicists were drawing all the time were truly reliable.

This might be surprising. After all, Feynman, Schwinger, and many others were regularly doing calculations with the theory, explaining all sorts of observed phenomena. Surely, you'd think, they must have had the mathematics worked out in order to do this. But in fact, physics develops in a much more roughshod way, involving a complicated mixture of mathematical calculation, physical intuition, and straight-up guesswork. Working physicists are more interested in the pragmatic issues of developing a feeling for how the world works and getting answers that are "good enough" to approximate experimental results than they are in dotting all the i's and crossing all the t's. This is especially true when physicists are working on a new theory. The mathematicians can come in later and clean up the details. Even relativity theory, which today is a model of clear mathematical physics, began life as a series of provocative thought experiments and somewhat dubious physical arguments.

Feynman, in particular, was an intuitive wizard: throughout his career, his approach to solving problems often seemed magical, with leaps from one idea to another that others could rarely follow—but which invariably led to the "right" answer, or at least, the answer that others would ultimately arrive at using more mundane methods. In fact, although physicists would ultimately adopt Feynman's pictorial approach to solving problems in quantum field theory over Schwinger's more traditional methods, Feynman's ideas were held in great suspicion during the 1940s.[167] At one conference in 1948, for instance, chaos erupted during Feynman's talk, with distinguished older physicists like Niels Bohr insisting that Feynman simply hadn't mastered basic concepts from quantum theory. Afterward, Feynman declared the lecture a complete failure because no one had understood a word he said.

Ironically, although he didn't understand what Feynman was doing any better than the others, Schwinger was far more willing to trust him, because Schwinger was one of the few people at the time who could independently check Feynman's results. Schwinger's approach to QED, meanwhile, was initially better received by the rest of the physics community. But although he was much more committed to using standard mathematical methods, the problems he encountered were extremely difficult, and he, too, sometimes resorted to intuitive leaps and guesswork.

Wightman was a graduate student at Princeton during the late 1940s as these controversies unfolded. He was working with John Wheeler, the same person who had supervised Feynman's work five years earlier. Wightman took it upon himself to fill in the gaps left by Feynman and Schwinger's more heuristic methods. His approach was to identify a handful of principles that Feynman and Schwinger were using to develop their theory, to codify them as precise mathematical assertions, and to try to prove the results that Schwinger, Feynman, and others had already achieved. The difference was that Wightman aimed to do this using pure mathematics. Since the theory had already had major successes and no one doubted that it was, at some level, mathematically correct, Wightman's project seemed manageable, if perhaps unnecessary for the purposes most physicists cared about.

It would be wrong to say that Wightman—and Haag and Kastler, and dozens of their students and collaborators—failed. They were able to establish many important results concerning quantum field theory and to launch an important research program. But neither has that program succeeded. More than sixty-five years after Wightman began working on the mathematical foundations of quantum field theory, mathematical physicists still have not developed a way of treating realistic quantum field theories, including QED, at the level of mathematical rigor that mathematicians demand.

This is not to say that no progress has been made. For instance, many of the most troubling problems from the early days of the theory, related to a procedure known as "renormalization" that allowed physicists to extract meaningful predictions from a theory that tended to give nonsensical answers, were resolved by the groundbreaking work of a physicist named Kenneth Wilson during the 1970s—work for which Wilson won the Nobel Prize in 1982. Wilson took what had been a handful of ad hoc and dubious maneuvers and showed how they could be given a coherent physical interpretation. But while Wilson shored up the most glaring cracks in the theoretical edifice, he did not solve all of the problems.[168] As Arthur Jaffe, one of Wightman's former students and a professor of physics and mathematics at Harvard, recently put it in a talk at UC Irvine, we still don't understand quantum field theory.

There are many reasons to find this surprising. The theory that Dirac, Feynman, and Schwinger established has been elaborated and expanded to form the theoretical basis for the Standard Model of Particle Physics, which was assembled from various pieces in the early 1970s. The Standard Model, meanwhile, has gotten forty years of experiments exactly right. It is an extremely powerful theoretical machinery.

There is no way that the Standard Model could have had these successes if physicists didn't, in the most important sense, understand quantum field theory. And yet no one has managed to clearly and completely articulate how the theory works using mathematics—even in light of Wilson's insights. Most physicists are not troubled by this. Feynman, in particular, dismissed worries about the mathematical rigor of quantum field theory. But Schwinger took these problems very seriously and worked for much of his career to put the theory on a more solid mathematical footing. And Dirac found the mature theory of QED that Feynman and Schwinger developed downright disturbing: it was "complicated and ugly," he said on one occasion; on another, he declared it was "just not sensible mathematics."[169]

Where does this leave the quantum vacuum? Actually, if *anything* in quantum field theory is well understood, mathematically speaking, it is the vacuum.[170] Some of the interpretations of vacuum phenomena that I noted above, concerning matter and antimatter and spontaneous fluctuations, are perhaps suspect. Even vacuum polarization, though well established experimentally, is not completely understood. But the basic idea that these interpretations were supposed to give us some handle on, concerning the probability of finding stuff in a universe with nothing, *has* been established, to the highest standards of rigor. There are many places where one might worry about the rigor of quantum field theory. But the strange features of the vacuum aren't among them.

Epilogue

Why Nothing Really Matters: Quantum Gravity and Beyond

In 1973 and 1974, particle physics was rapidly transformed.[171] A series of theoretical and experimental discoveries definitively established that the particles in the nucleus of an atom—protons and neutrons—are themselves composed of another kind of particle, called *quarks*. During the same period, a quantum field theory of quarks was proposed that could explain essentially all known experimental results concerning protons, neutrons, and many other particles—known collectively as *hadrons*—which, it turned out, were also composed of quarks. This new theory was so similar to quantum electrodynamics that it is now called quantum chromodynamics, or QCD for short. (The "chromo-" refers to the fact that quarks carry a new kind of charge, often called color charge.) Along with a close descendant of QED, QCD is a core part of the Standard Model of Particle Physics. The great successes of QCD and the quark model, which had been around for about a decade, independently of QCD, led to a number of big winners: in those two years, five physicists received Nobel Prizes for work related to quarks and QCD.[172]

But amid the fanfare, other theories that had been plausible contenders to explain hadronic physics fell by the wayside. One alternative theory, which had earned a small but devoted following during the 1960s, supplemented the quark models with tiny *strings*, which were supposed to account for how the quarks stuck together inside a hadron.[173] The theory had some successes and also some problems, but it did not last long once QCD broke onto the scene.

In 1974, a few former devotees of the now-defunct string theory—Joël Scherk and John Schwarz, and (independently) Tamiaki Yoneya—decided to dust it off, give it a new coat of paint, and turn it around as a new kind of theory. In this second incarnation, however, string theory was no longer a way of explaining what kept quarks together. The new vision was much more ambitious. Now the strings were supposed to be far smaller than in the old theory—as much as 10^{20} times smaller—and they were no longer the ribbons tying up packages of quarks. In the new theory, strings were at the very foundations of physics. String theory was proposed as a theory of everything.

The theories I have discussed in this book—Newtonian mechanics, Maxwell's electromagnetism, relativity theory, quantum field theory—present radically different conceptions not only of what the world would be like if there were no stuff around, but also of what the word *nothing* even means. In Newtonian physics, a world with nothing would consist in empty space and time. In such a universe, space-time has a fixed, immutable geometrical structure whose basic properties would not change if stuff were suddenly to appear. It is simply a container standing ready for stuff to be introduced.

In general relativity, a world with nothing would again be a universe devoid of any stuff—that is, space-time with no matter present in any location. But far from fixed and immutable, the geometry of space-time is rich and dynamic, even without any matter at all. And this richness allows purely geometrical phenomena that have many of the hallmarks of stuff: black holes that affect nearby observers in spaceships just as planets or stars do, or gravitational waves that set springs vibrating, light lightbulbs, and even knock your socks off. The geometry of empty space-time is such that few of our expectations about what is possible in a universe with nothing are met.

We also saw, in connection with the discussion of general relativity, that Maxwell's theory of the electromagnetic field not only introduced a new kind of stuff into physics—the electromagnetic field—but it also raised significant questions about what it would mean for a region of the universe, or even the universe as a whole, to be empty. If electromagnetic radiation, including light, is to count as a kind of "stuff"—as it must in general relativity, since it contributes to the curvature of space-time in the same way that the sun does—then a region of space and time that is empty of everything except for light (say) is not empty at all. More importantly, I raised the question of whether a universe with no vibrations, or excitations, in the electromagnetic field should count as a universe with no field at all. Is light the stuff, or is light just a particular pattern in a field that, like an aether, pervades the entire universe whether it is vibrating or not?

This last question became especially pressing when we moved to quantum field theory. There the state of affairs with "nothing" in the universe is represented by a vacuum state. But this sort of state does not have the properties we might expect of a universe with nothing in it: in particular, if you start looking around, you may well find some particles. This vacuum state, I suggested, corresponds closely to a situation in which there are no excitations in the electromagnetic field—but where there is, nonetheless, an entity (the field) present, which can give rise to observable phenomena, such as vacuum polarization. I described some heuristic ways of understanding this—vacuum fluctuations, electron/positron creation and annihilation—but the basic facts are about what sorts of measurement outcomes are possible in an empty universe, and those are simply bizarre.

This bestiary of "nothings" has a few features that stand out. One is that in each of these theories, there is a perfectly precise sense—more precise, in fact, than in our ordinary thinking about nothingness—of what it would mean for there to be nothing: empty space-times, zero fields, vacuum states. These precise notions of nothingness are not unrelated to our ordinary concept of nothing as the lack of "something," but they aren't identical to it, either. Instead, in each case, in the course of developing a physical theory, physicists have taken the ordinary notion of nothingness (informed, of course, by previous physical theories) and honed it, spruced it up, and put it to work.

This kind of move is ubiquitous in science. Ideas that have currency in day-to-day life—"species," "particle," "energy," "bond," "fitness," "life," "randomness"—get transformed as they are put to work for specific scientific purposes. The precision of scientific theorizing requires us to be rigorous about the meanings of our terms. But this very precision also creates new possibilities for confusion—and for conflict. We often want to bring our best science to bear on questions of philosophical, political, or even religious significance. But in doing so, we need to recognize how we often change the subject, shifting between the ordinary senses of words and the more specific, nuanced, and subtle senses these words have in science.

To make the point, consider again the question I raised in the introduction to this book. *Why is there something rather than nothing?* As I suggested, one might think that this is the sort of question that will ultimately be answered by our best physics. But we can now see just how much the question's very meaning depends on the details of a background theory of physics. Perhaps the most extreme case is that of the vacuum state in quantum field theory. There, "something" and "nothing" are not even mutually exclusive: it is entirely consistent for there to *be* nothing, and nonetheless for stuff to show up when we try to detect it. But even in general relativity, there are variants on nothingness that seem an awful lot like stuff. Why, then, are there black holes, rather than nothing? It turns out that black holes are just a variant of nothingness.

Are these sorts of answers satisfying? They are certainly not responsive to the original question: by "something," one likely did not mean "gravitational wave" or "vacuum fluctuation," and by "nothing," "quantum vacuum state." But what *did* we mean? Did we, perhaps, have in mind some implicit "theory" of what the world would have been like if there were nothing, and conversely, of what sorts of things count as "stuff"? If so, then it seems significant that the ways in which these ideas have been co-opted by our best physics do not conform to those prior intuitions. If our previous understanding relied on some sort of theory, is it as well supported as quantum field theory or general relativity?

And if not, shouldn't we rethink why we cared about the question in the first place?

The different ways in which our ordinary concepts can be repurposed for scientific work do more than merely engender confusion when brought to bear on pretheoretical, or traditionally "philosophical" questions. They can also create conflicts within science. Whatever else is the case, "nothing" in general relativity has a completely different meaning than "nothing" in quantum field theory. And there is good reason to think that *both* senses of "nothing" will be left behind in future physical theories.

Let me explain. The huge leaps forward in particle physics during the 1970s brought three "fundamental forces" under control, within the largely unified theoretical framework of quantum field theory and the Standard Model. These three forces are the strong force, which keeps quarks bound together in protons, neutrons, and other hadrons and which is described by QCD; electromagnetism, essentially described by QED; and a force known as the "weak force," which explains some of the processes we see in the nucleus of an atom and that turned out to be another very close cousin of electromagnetism.[174] (In fact, QED has been superseded by a theory that combines electromagnetism and the weak force, known as "electroweak theory"; QED itself turns out to be a very good approximation to electroweak theory if one is interested only in electromagnetic phenomena.) In all of these cases, "nothing" means the quantum field theoretic vacuum—replete with fluctuations and creation/annihilation of many different kinds.

The Standard Model, with these three pieces, was complete in 1974, and except for filling in some minor details with ever-more-precise experimental results, it has remained unchanged since then. But physicists' successes in bringing these three forces under control only drew more attention to the fourth and only remaining "force," gravity. By the early 1980s, once the Standard Model had been fully articulated, particle physicists turned en masse to the problem of constructing an adequate theory of quantum gravity.[175]

This problem has turned out to be extremely difficult. Even understanding the goal is subtle. Following the usage of many particle physicists and quantum gravity experts, for instance, I just described gravitation as the fourth "force." This is a tempting way of thinking: we tend to use forces to explain motion, and gravitation is responsible for many of the apparent motions we actually see in the world, from cranes falling to the sun rising and setting. Moreover, in classical Newtonian physics, gravitation *was* conceived as a force, and in that context, it seems very strongly analogous to the three forces. So it seems like a natural target for attacking with the same basic tools used to produce quantum theories of the other three forces.

But in general relativity, these analogies begin to break down. It becomes far more contentious to think of gravitation as a "force." Recall that in general relativity, gravitational effects are just a manifestation of space-time curvature. A force, in both Newtonian physics and in relativity theory, is something that accelerates a body—that is, something that makes the body deviate from straightest-line motion. But in general relativity, gravitation doesn't do that. So gravitation isn't a force, at least not in the same way that electromagnetism and the strong and weak forces are. This makes it very difficult to see how to bring gravity—or at least, general relativity—under the control of quantum field theory.

This brings us to a serious tension in how the two theories—quantum field theory and general relativity—treat nothingness. In quantum field theory, the vacuum state encodes information about the probability of finding stuff if you make a measurement of the entire universe, over all time—and also, the probability of finding anything if you looked in smaller regions, like your backyard or the space between earth and Mars. And as we noted, the probabilities of finding stuff in these smaller regions may be nonzero, even in the vacuum state.

All of this makes sense in the presence of some fixed background space-time structure that allows you to associate regions of the universe with possible observables. But in general relativity, you simply don't have this. The structure of space-time is not fixed once and for all. This means there is no way to speak of "a region" of space-time once and for all, even in the absence of any stuff.[176] To make matters worse, the structure of space-time actually depends on whether stuff is present, and what kind of stuff. So if you try to fix an empty space-time, pick a region of that space-time, and then go and measure the particles in that region, you're going to run into problems. This is because there is a possibility that you would find particles there after all—which means, in turn, that the geometry of space-time couldn't have been what you originally thought, since you began by assuming space-time was empty!

I do not mean to suggest that there is some fundamental inconsistency with the idea of quantum gravity, or even with a quantum theory of general relativity. But there are real challenges, many of which come down to incompatibilities between basic concepts of general relativity and quantum field theory. "Nothing" may be a bit weird in both theories, but in each case alone it is coherent. Put them together, however, and things start breaking down. Figuring out how to make these very different theories play nicely together is one of the most difficult and important questions facing modern physics—and it has been for more than four decades.

Physicists have long been aware of the many difficulties associated with finding a theory that weds general relativity and quantum field theory. Efforts to develop a quantum theory of gravity are as old as QED, with initial attempts dating back to the 1930s.[177] During the 1950s, Dirac and others initiated a program now known as "canonical" quantum gravity. ("Canonical" because the strategy drew on something known as "canonical variables" in classical physics, not because this method is, or ever was, canonical!)

This program went about as far as it could by the mid-1960s, with the work of Wheeler (Feynman's and Wightman's PhD supervisor) and Bryce DeWitt, who had been a student of Schwinger's at Harvard. Together, they produced what is now known as the Wheeler-DeWitt equation, an analogue of the Schrödinger equation for a quantum theory of gravity. Ultimately, though, no one could figure out how to use the Wheeler-DeWitt equation to solve any real problems, and the approach floundered. One might think of canonical quantum gravity as a "brute force" approach to the problem, and at some point it became clear that more subtlety was needed.

The main selling point of string theory, in its new incarnation as a theory of everything, was that it provided the required subtlety. The idea behind the theory was that, at root, the world consists of tiny vibrating strings, whose different possible vibrations corresponded to different possible particles, with different masses, charges, and so on. One of the major problems facing the *old* string theory was that it appeared to offer too much: in addition to the hadrons it was supposed to describe, the old string theory required the existence of a new kind of particle that no one had observed. This wasn't fatal, but it was certainly a concern.

When the theory was repurposed by Scherk and Schwartz, however, this strange, problem particle went from being an embarrassment to being the theory's biggest virtue. They argued that this particle had precisely the properties that one would expect in a *graviton*—that is, a particle associated with a quantum theory of gravity in the way that the photon is associated with the electromagnetic field in QED. In other words, string theory appeared to have a quantum theory of gravity baked right in, with no need to "quantize" general relativity directly.

At first, few physicists bought into the new theory. But in the early 1980s, that changed. On the one hand, Schwarz and a young collaborator named Michael Green showed that string theory managed to avoid a certain class of problems, known as "anomalies," that plagued other similar theories. On the other hand, it became increasingly clear how difficult it would be to develop a quantum theory of gravity. String theory suddenly became very attractive, and a whole generation of theorists began to work on it.

The theory had some early successes—but also some problems. For instance, it appeared to require that the universe was not three- or four-dimensional, as borne out by ordinary experience and as taken for granted by most other physical theories, but rather *ten*-dimensional, where the six "extra" dimensions were wound up very tightly, so that they couldn't be readily observed. It also proved very difficult to extract meaningful predictions from the theory—at least of a sort that could be tested with contemporary technology. For these reasons, many leading physicists of the previous few generations objected to the theory. Feynman, for instance, wrote in 1988 that string theory "doesn't produce anything, it has to be excused most of the time. It doesn't look right."[178]

For the younger physicists attracted to the theory, these problems were just growing pains—and the objections of Feynman and others were merely caviling by old-timers who hadn't bothered to learn the new theory. After all, Einstein never accepted the new quantum theory, and if that didn't stop Feynman, why should Feynman stop the string theorists?[179]

One of the earliest goals of string theory was to show that by properly analyzing the possible configurations of strings, one would be able to show that the Standard Model is, in some sense, *inevitable* within a universe governed by string theory.[180] That is, string theorists hoped to show that the physics we see at particle accelerators is somehow implied by this new theory, properly construed. If this could be shown, the theorists believed, it would both give some credibility to string theory and also provide a kind of explanation for why the world is the way it is: it is the only way it *could* be, given that it is constructed out of strings.

As the '80s wore on, however, this goal became ever more elusive. No one was exactly sure how to get from string theory to the Standard Model or anything like it—never mind to general relativity. But there were some unsettling signs that even if string theory could lead to the Standard Model, it was unlikely to do so uniquely.

Then, in 1987, a group of theorists in Switzerland and Germany published a paper in which they made a shocking conjecture.[181] Not only was the Standard Model not *uniquely* implied by string theory (if, indeed, string theory was compatible with the Standard Model at all!), but in fact there were *many* alternatives to the Standard Model that would be equally compatible with string theory. By their estimate, the number was approximately 10^{1500}. (Today, this number has been revised downward. Most estimates put the number at 10^{500}—though the difference hardly matters.[182]) By contrast, there are believed to be well under 10^{100} atoms in the visible universe.

These estimates are striking for all sorts of reasons. But the one that is most important for our purposes is that each of these possible alternatives to the Standard Model corresponds to a distinct *vacuum state* allowed by string theory. In other words, what string theorists gradually realized in the late 1980s is that string theory allows for approximately 10^{500} *different* ways for there to be nothing.

What does this mean? On the one hand, it is not so surprising: after all, general relativity arguably allows for *infinitely* many ways for there to be nothing in the universe, since the theory allows infinitely many empty universes. But the string theory result has a different character. Recall that in quantum field theory, one way of understanding the strange properties of the vacuum is that "fluctuations" can produce particles that live for a short time and then disappear. But what sorts of fluctuations are possible? It depends on what kind of vacuum state there is. The vacuum state in QED, for instance, allows virtual electron/positron pairs to appear and disappear.

But QED doesn't say anything at all about, for example, quarks popping into existence. This is because the vacuum state in quantum field theory is just a state of some *particular collection of fields*, like the zero state of the electromagnetic field. The fluctuations that are possible in a given vacuum state depend on what fields that state characterizes. QED doesn't say anything about quarks, and so the QED vacuum doesn't include any quark fields that could fluctuate.

This is a good way of understanding the radical nonuniqueness of the string vacuum, too—and also of understanding why this bears on whether string theory uniquely gives rise to the Standard Model. When physicists say that there are 10^{500} possible vacuum states in string theory, this is a way of saying that string theory appears to allow for 10^{500} possible combinations of fields, each combination corresponding to some distinct zero state. The Standard Model describes just one such possible combination.

The vast number of string vacua didn't spell doom for the theory, but it did lead to puzzles. Far from explaining why the universe is the way it is, string theory appears to tell us that there are far, far more possibilities than we ever considered. If all of these vacua are possible, why do we happen to see a world governed by the Standard Model?

Such questions have gradually led theorists to reconsider what string theory is trying to accomplish. One prominent idea that has emerged in response to these puzzles is due to Stanford physicist Leonard Susskind.[183] Susskind has suggested that in fact *all* of the possible string vacua are realized in nature. One can think of this in several ways: either the universe is much vaster than we thought, and different regions of it correspond to different string vacua; or else there isn't *one* universe at all but many different ones, each corresponding to a different vacuum. Such theories are known as "string landscape" or sometimes "multiverse" theories.

Broadly, this approach to string theory changes the subject: rather than explain why the universe *has* the properties it has, one tries to explain why the universe we *see* has the particular properties it does. The idea is that if there are many, many universes—say 10^{500}—but we couldn't exist in the vast majority of them, then it isn't surprising that we happen to find ourselves in a corner of the multiverse where physics works in a way compatible with our existence.

The multiverse is an extremely controversial idea. For some, it is a natural consequence of the most compelling theory known to fundamental physics. To others, it is an abdication of responsible science—a theory that makes no predictions and places no constraints on what the world could possibly be like. This is not the occasion to weigh in on such matters. Suffice it to say that, whatever one thinks of the viability of the string landscape, it is a landscape of different kinds of nothingness. Once again, the physics of nothing is front and center, and we are forced to reckon with a new conception of what the world would be like if there weren't anything at all.

Acknowledgments

I am grateful to many friends and colleagues for their help and support while I was writing this book. From the earliest stages of the project, I benefited from helpful conversations and correspondence with Jeremy Butterfield, Tony Duncan, John Earman, Ben Feintzeig, Peter Galison, Bob Geroch, Marian Gilton, Michel Janssen, Pen Maddy, David Malament, John Manchak, John Norton, Sarita Rosenstock, Howard Stein, and Giovanni Valente; my views on these topics have also been deeply influenced by discussions with (and the written work of) Gordon Belot, Harvey Brown, Erik Curiel, Robert DiSalle, Hans Halvorson, Sam Fletcher, Eleanor Knox, Oliver Pooley, Chris Smeenk, and Chris Wüthrich. I am especially grateful to Tony Duncan, Jeff McDonough, Chris Smeenk, and John Norton, who were kind enough to carefully read early drafts of various sections of the book and alert me to infelicities; and to Jeff Barrett, Erik Curiel, Sam Fletcher, John Manchak, David Malament, Cailin O'Connor, and two anonymous referees, who read the entire manuscript in draft form and who gave me invaluable advice. Of course, remaining errors are my own.

I am also grateful to my agent, Zoë Pagnamenta, for her support over the years and for helping bring this book into existence, and to Bill Frucht, my editor at Yale University Press, for his belief that a book of this nature could and should be written, and for his help in preparing the final manuscript.

Most of this book was written while I was a visiting fellow at the Center for Philosophy of Science at the University of Pittsburgh. I am grateful for the support of the center. Parts of Chapters 1 and 2 evolved out of lecture notes written for a course at UC Irvine titled, "What Is Time?," which was developed with the support of the National Endowment for the Humanities under Grant No. AQ51039; more generally, the material here is based upon work supported by the National Science Foundation under Grant No. 1331126. I am also grateful to the School of Social Sciences at UC Irvine for considerable research support during the period that I was writing.

Finally, I would like to thank my parents, Maureen and Jim Weatherall; my wife, Cailin O'Connor; and my daughters, Eve and Vera Weatherall, for their love and unending support. Without them, this would never have been written.

Notes

PROLOGUE

1. This question was recently taken up in two books, by Krauss (2012) and Holt (2012). Holt's book consists of interviews with physicists, philosophers, and others, attempting to tackle the question from as broad a perspective as possible. It is remarkable. Krauss, meanwhile, describes a speculative proposal concerning the origin of the universe and then argues that this proposal has significance for religious belief. In a scathing review of the book, David Albert (2012) criticized the sense in which "nothing" is used in Krauss's book, pointing out that "nothing" as understood in modern physics is strikingly different from "nothing" as understood pretheoretically. This review led to a heated exchange, mostly on the Internet(!), with a number of prominent physicists and philosophers weighing in on everything from what Krauss had accomplished to whether philosophy has any value. I have no intention of wading into these battles here. But the exchange was, in a roundabout way, the inspiration for this book. Albert is certainly right about one thing: "nothing" as conceived in quantum field theory is radically different from our pretheoretic intuitions—and even from other theories of physics. My goal here is just to expand on Albert's point, to give an appreciation of how a concept like "nothing" can be appropriated by physical theories and how such appropriation can change how we understand basic questions about the world. My own view is that this sort of reappropriation is often quite fruitful, and the versions of questions that we find *within* the sciences are the more important ones. In this way, I am sympathetic with Krauss.
2. In a sense, the question understood in these stark terms bears some resemblance to the question behind radical skepticism: Can we justify our beliefs about the world without appealing to any of those beliefs? See Stroud (1984) and Maddy (2007) for a discussion for this question.

3. In addition to the references in note 1, see Barrow (2001) and Webb (2013) for discussions of the physics of nothing aimed at a general audience.
4. See note 12 for some caveats about Newton and the "container" view of space. It is worth noting that Newton himself distinguished empty space from "nothing" in the purest sense, because he took space to be an aspect of creation. See Stein (2002) for a discussion.
5. As I explain in chapter 3, the "old" quantum theory was developed beginning around 1900; it was the "new" quantum theory that was developed in the 1920s. (See note 114.) Quantum field theory, meanwhile, was first developed in the 1920s, around the same time as the new quantum theory. But the modern form of the theory was not produced until the late 1940s.
6. For other kinds of claims to the effect that modern physics answers the question, see Tryon (1973) and the references in notes 1 and 3.

CHAPTER 1

7. Several good biographies of Newton exist, including Westfall (1980) and Hall (1992). Briefer, but reliable, overviews of Newton's life and major works can be found in Iliffe (2007) and Smith (2008). For more on the broader cultural influence of Newton's work in physics and elsewhere, see Cohen (1983), Feingold (2004), and Buchwald and Feingold (2012). Many of Newton's most important papers, with commentary from leading Newton scholars, can be found in Cohen and Westfall (1995).
8. The classic history of calculus, which discusses Newton's contributions (and Leibniz's) and also puts their ideas in historical context, is Boyer (1959). See also Kline (1972). For histories of the priority dispute between Newton and Leibniz, see Westfall (1980, pp. 698–780), Meli (1993), and Hall (2002a); a shorter overview is given in Hall (2002b). For more on Newton's optics, see the texts and commentary in Cohen and Westfall (1995), as well as Sabra (1981), Shapiro (2002; 2013), and references therein. Connections between Newton's work in optics and some of the other methodological themes discussed here are given by Stein ("On Metaphysics and Method in Newton," unpublishe) and "Further Considerations on Newton's Methods," unpublished).
9. My understanding of Newton's views on space and time and of the role that these views played in his larger project is heavily indebted to Howard Stein—especially Stein (1967; 1977; 2002; "On Metaphysics and Method in Newton," unpublished)—and Robert DiSalle—especially DiSalle (2002; 2006). In particular, the picture I present here is one on which Newton is not especially interested in the sorts of metaphysical questions concerning substantival and relational conceptions of space that have often been associated with him (especially in the context of his dispute with Leibniz, as described in this chapter). Rather, he is interested in identifying and characterizing the spatio-temporal structure needed to support his laws of motion. Leibniz, of

course, *is* interested in more purely metaphysical issues—and so is Samuel Clarke, Newton's amanuensis, also discussed in this chapter—which partially explains why their disagreements have been interpreted as they have been. For a more traditional reading of Newton's views, see Rynasiewicz (1995a; 1995b); for an explicit argument *against* DiSalle's reading of Newton (and Stein!), see Huggett (2012). Janiak (2008) offers yet another take on the role of metaphysics in Newton's thought; for a detailed discussion of the relationship between Steinian readings and Janiak's views, see Domski (2010). A detailed treatment of the substantivalism/relationism debate, from Newton and Leibniz to the twentieth century, is given by Earman (1989).

10. The authoritative translation of the *Principia* is that of I. Bernard Cohen and Anne Whitman, based on the third (1727) edition (Newton, 1999 [1687/1713/1726]). For background and guidance on reading the *Principia*, see Cohen's guide, published along with the translation (Cohen, 1999); see also Smeenk and Schliesser (2013). Discussions of Newton's methods in the *Principia* and elsewhere, such as those given by Smith (2002), Harper (2012), and Stein ("On Metaphysics and Method in Newton," unpublished), are also helpful. *De Grav* was first published as Newton (1962). The most important commentary is Stein (1967), though see also Stein (2002 and "On Metaphysics and Method in Newton," unpublished; the second of these has some important notes on the standard translation) and DiSalle (1994; 2002; 2006). There is no consensus on when *De Grav* was actually written. Many scholars, including the Halls, in their original commentary on the paper, hold that it dates to an early period, and perhaps even the 1666–1667 plague year; others, perhaps most notably Dobbs (1991), argue that it is a much more mature work, dating to the period in late 1684 when Newton was writing the *Principia*. A recent overview of the issues, including a review of the literature on *De Grav* and some compelling evidence against the Dobbs reading, is given by Raffner (2012).
11. The language of "place" and "space" comes from Newton's Scholium to the Definitions at the beginning of Book 1 of the *Principia* (Newton, 1999 [1687/1713/1726], pp. 408–415). See note 39 and the surrounding discussion.
12. The "container" picture of Newtonian space is potentially misleading in several ways. For instance, it suggests that space is (ontologically) "prior" to and independent of bodies. It is not clear that Newton believed this, at least in the strongest sense that some commentators attribute to him. For instance, in *De Grav* (as translated by Stein), Newton says that space is an "an emanative effect of the first-existent being, for if I posit any being whatsoever space is posited" (Stein, "On Metaphysics and Method in Newton," unpublished, p. 32). I take this (particularly the expression "emanative effect") to mean that although existence *in space* is necessary for existence of any being (i.e., body, mind, or God), the existence of space follows from the existence of bodies.

13. Helpful resources on historical atomism and its evolution, including many references, are given by Berryman (2011) and Chalmers (2014).
14. Perhaps the elementary particles noted above—electrons, quarks, etc.—are closer analogues to the "atoms" of the seventeenth century and before than are atoms in the contemporary sense. Even so, the particles of modern particle physics, as described by quantum field theory, bear so little resemblance to particles or atoms in the classical sense that the analogy is simply unhelpful.
15. For more on Newton's atomism, see Shapiro (2002) and Chalmers (2014).
16. For background on conceptions of empty space in the Middle Ages, up to and including the seventeenth century, see Grant (1981) and Sorabji (1988). Jammer (1993) is another classic, providing a history of theories of space from the Ancients through the twentieth century. Aristotle's arguments against the possibility of a vacuum or void are given in Book 4, Sections 1-9 of the *Physics* (Aristotle, 1984). The argument I attribute to Aristotle in the text is meant to capture the spirit of his views, though in fact, he offered a number of different and subtle arguments against the existence of a void; these are reviewed, with a discussion of how other Ancients replied, by Sorabji (1988, pp. 142-159).
17. Again, see Grant (1981) and Sorabji (1988) for detailed histories of discussions of empty space during the Middle Ages, leading up to Newton.
18. Descartes's most mature and complete attempt to provide a theory of physics appears in his *Principles of Philosophy* (Descartes, 1985). For an overview of Descartes's physics, see Slowik (2014) and references therein; for more detail, see Garber (1992).
19. This argument appears in II.16 of *Principles of Philosophy* (Descartes, 1985, pp. 229-230).
20. At least, his views were rejected by Newton and Leibniz, though many Continental physicists remained Cartesians.
21. The passage from the letter appears in Turnbull (1959, p. 416) and is quoted in Westfall (1980, p. 274). Though the reading I suggest here is amusing, Westfall and others reject it—though Westfall, at least, recognizes a "lack of warmth" in Newton's professions of praise of Hooke.
22. The definitive biography of Leibniz is Antognazza (2009); an in-depth overview of his philosophical work is given by Jolley (2005). Many of Leibniz's own writings are collected together in Leibniz (1989). Shorter and more accessible overviews of his life and works can be found in Ariew (1995) and Look (2014). For details specifically on Leibniz's physics (and philosophy of physics), see Garber (1995) and McDonough (2014); see also Garber (2009) for a more in-depth treatment of how Leibniz's physics and metaphysics are related to one another. As with Newton, my understanding of Leibniz's views on space, time, and motion are again heavily indebted to Howard Stein—par-

ticularly Stein (1967; 1977)—though it is due to Daniel Garber's influence that the word "monad" appears only here!

23. See Antognazza (2009, pp. 226–227); chapter 5 of Antognazza details the two-and-a-half-year journey Leibniz took through South-Central Europe to document the Hanoverian genealogy!
24. A clear description of the tension between Leibniz's formal education and his personal studies is given by Antognazza (2009, pp. 30-37).
25. For more on Leibniz's "conciliatory eclecticism" and its role in his intellectual development, see Mercer (2004); this aspect of Leibniz's intellectual personality is also emphasized by Antognazza (2009). I am grateful to Jeffrey McDonough for help on this matter. The story of Leibniz's forays into alchemy is from Ariew (1995, p. 21); for more on Newton's alchemy, see Dobbs (1991) and Figala (2002).
26. Many of Leibniz's (critical) writings on Descartes are collected in Leibniz (1989, pp. 235–267). Leibniz's views on Aristotle and Descartes are treated concisely by Jolley (2005); see also Mercer and Sleigh (1995). For an example of Leibniz attempting to reconcile a Cartesian doctrine with Aristotelianism, see Garber's (2009, pp. 7–9) discussion of Leibniz on Aristotle on mechanistic philosophy.
27. For more on Leibniz's views on empty space, see Garber (1995, pp. 304–305; 2009, pp. 44–45n145) and section 5.2 of McDonough (2014).
28. Although Newton's work was recognized as important and widely studied, the history of its reception is actually somewhat complicated, and it seems many of his contemporaries, particularly on the Continent—including Leibniz—were either unable to understand what he had accomplished in the *Principia* or else felt that he had accomplished *less* than Descartes because he had not given a causal/mechanical account of gravitation. See the final section of Smeenk and Schliesser (2013) for an enlightening discussion of the reception of the *Principia*. I am grateful to Chris Smeenk for correcting me on this point.
29. Once again, for histories of the priority dispute, see the references in note 8.
30. The book was *Analyse des Infiniment Petits, Pour L'intelligence des Lignes Courbes* (*Infinitesimal Analysis for Understanding Curves*; l'Hôpital, 1696). Though it was published anonymously, there is no doubt of its authorship.
31. The story of this review is given by, for instance, Hall (2002a, pp. 138–141) and Westfall (1980, p. 715). That Leibniz was the author of the anonymous review is certain: the editor of *Acta Eruditorum* identified Leibniz as author in his editor's copy (Newton, 1981, p. 26). Moreover, Westfall claims Newton never doubted that Leibniz was the author (Westfall, 1980, p. 715).
32. Newton's mental health problems in 1693 are detailed by Manuel (1968, pp. 213–225) and Westfall (1980, pp. 535-540); the suggestion that mercury was responsible is due to Johnson and Wolbarsht (1979) and Spargo and Pounds (1979).

33. The claim that the substance of Newton's creative work was finished by 1693 follows Westfall (1980, p. 540). Though Newton certainly continued to publish, most of what appeared reflected work he had done earlier in his life.
34. The historical background to the Leibniz-Clarke correspondence is given by Ariew (2000) and Meli (2002).
35. The correspondence has been translated and published, with commentary, as Leibniz and Clarke (2000 [1715–16]).
36. Ariew (2000) and Meli (2002) discuss, in detail, the evidence concerning Newton's role in writing the letters. See also Cohen and Koyré (1962) and Westfall (1980, p. 778).
37. Putting the assertion in this way is anachronistic—and it is likely that Leibniz would not have liked it, as he did not think that extended bodies were "composed" of points; rather, points were to be understood as secondary to lines or other continuous things. But I think it is harmless for present purposes. I am grateful to Jeff McDonough for help on this point.
38. In fact, Newton gives an argument very much like this one to establish that space is infinite in *De Grav*. See Newton (1962, pp. 133–134). I am grateful to Chris Smeenk for drawing my attention to this passage.
39. Newton draws these distinctions in his famous Scholium to the Definitions at the beginning of Book 1 of the *Principia* (Newton, 1999 [1687/1713/1726], pp. 408–415). To understand what Newton had in mind, it is useful to note that he actually distinguished "absolute, true, and mathematical" space, time, place, and motion from "relative, apparent, and common" space, time, place, and motion. The first set of notions were the ones represented by the mathematics of the *Principia*; they were only approximated by the second set of notions, which are what we directly measure. There is a very large philosophical literature on absolute and relational theories of space. Earman (1989) is the modern classic on this topic; see also Huggett and Hoefer (2015) and references therein.
40. Though Leibniz certainly does attack Newton's views on absolute space on metaphysical grounds, there is some reason to believe that he simply did not understand what Newton's views really were—and in particular, that Leibniz conflated "absolute space" with the idea that space must be a "substance" in something like the Aristotelian (or Leibnizian) sense. For arguments that Newton did not have that view at all, based mostly on *De Grav*, see DiSalle (1994; 2002; 2006) and Stein (2002; "On Metaphysics and Method in Newton," unpublished).
41. Leibniz gives essentially this argument in paragraph 5 of his third letter (Leibniz & Clarke, 2000 [1715–16], pp. 14–15). (In fact, Clarke first gives the argument, and Leibniz repurposes it in his reply. Thanks to Erik Curiel for reminding me of this!) For further discussion of Leibniz's arguments, see section 5.1 of McDonough (2014) and references therein. See also Pooley (2013) and Weatherall (2016b) for discussions of this argument in a more modern context.

42. For more on Newton's metaphysics of space, and how his metaphysical claims are deeply entwined with methodological issues regarding what structure he needs to posit to support the sorts of physics he goes on to do, see Stein (2002; "On Metaphysics and Method in Newton," unpublished; "Further Considerations on Newton's Methods, unpublished; 1977).
43. Clarke moves from absolute space to absolute motion in paragraph 4 of his third letter (Leibniz & Clarke, 2000 [1715–16], p. 19) and then cites Newton's example in paragraph 13 of his fourth letter (Leibniz & Clarke, 2000 [1715–16], p. 31). See also note 48.
44. For a fascinating and informative discussion of how Newton's conception of force is related to Leibniz's—and, relatedly, how their conceptions of "absolute motion" differ—see Garber (2012).
45. Note that in fact all of these bodies are *rotating*. There is a fascinating issue, discussed in detail by DiSalle (2006), concerning the status of nonrotational, uniformly accelerated motion. According to Corollary VI to the laws of motion in the *Principia*, uniform linear acceleration is indistinguishable from unaccelerated motion—suggesting, perhaps, that we could never say whether a body is really nonaccelerating, or if it is merely undergoing uniform linear acceleration, along with all of the reference bodies used to determine its relative motion. But, as I will explain below, this does *not* apply for rotational motion, and so it does not apply to the examples discussed here. For more on the significance of Corollary VI for the structure of space-time, see Saunders (2013), Knox (2014), and Weatherall (2016a).
46. Well, perhaps not *entirely* hopeless. See, for instance, Barbour and Bertotti (1982). Thanks to Erik Curiel for twisting my arm into adding this reference!
47. What Leibniz seemed to want was a sense of absolute motion that was intrinsic to a body so that whether something was in fact moving had nothing to do with motion relative to absolute space (or any reference body), but was rather just an internal state. But this leads to immediate problems, since it places no constraints on the relative motions of bodies that are undergoing absolute motion. For instance, in principle one could have two bodies that are at rest relative to one another, but only one of which is moving absolutely. See Garber (2012) and McDonough (2014), as well as Stein (1977).
48. Clarke himself points out that the argument can be run in this way in paragraph 4 of his third letter (Leibniz & Clarke, 2000 [1715–16], p. 19), and in reply, Leibniz *does* run the same argument against absolute motion that he runs against absolute space, claiming that the very idea of putting the whole world in motion along a straight line is incoherent (Leibniz & Clarke, 2000 [1715–16], pp. 23–24). It is telling, here, that Clarke moves directly from absolute space to absolute motion in responding to Leibniz's earlier arguments; it is in connection with this move, and Leibniz's reply, that Clarke cites Newton's early discussion of rotation in the *Principia* (Leibniz & Clarke, 2000 [1715–16], p. 31).
49. Newton's bucket example appears in the Scholium to the Definitions (New-

ton, 1999 [1687/1713/1726], pp. 412–413). Clarke actually points Leibniz to the eighth (and final) definition, concerning centripetal force, which appears immediately before the Scholium, but which is essential for the water bucket example.

50. The article is Stein (1967). See note 9 for more on Stein's work.
51. The view I am attributing to Leibniz here is a corollary of the idea that all motion is relative as I understand it. But Leibniz's views on this subject are complicated, and it seems that he *did* believe that there is a fact of the matter about whether a body is *forced*, and, arguably, whether it was moving, so it is possible (likely!) that Leibniz would not be happy with this way of putting things. See note 47 for further discussion. I am grateful to Jeffrey McDonough for help on this point.
52. Example offered with apologies to J. B. Manchak.
53. In fact, this structure had been described previously by Weyl (1922). Stein's contribution was to show how Newton's discussion of absolute motion can be understood in this four-dimensional framework. (I am grateful to David Malament for pointing out the relationship between Stein's work and Weyl's much earlier work.) Whether it is *precisely* this structure that is necessary for Newton's laws has recently been questioned by Saunders (2013). See note 45 for a discussion of the relevant issues.
54. Once again, see notes 47 and 51.
55. Voltaire comments in *Élements de la Philosophie de Neuton* (Voltaire, 1992 [1738]), du Châtelet does so in *Institutions de Physique* (du Châtelet, 1988 [1742]), and Euler does in a paper titled, "Reflexions sur l'espace et le temps" (Euler, 1750).

CHAPTER 2

56. Isaacson (2007) is an excellent, and eminently readable, biography of Einstein; I rely on him for many of the biographical details here. Pais (1982) is also an authoritative classic. Other good, if sometimes idiosyncratic, biographies are given by Frank (1947), Hoffman (1972), Bernstein (1973), and Neffe (2009). See also Einstein's autobiographical remarks, published as Einstein (1949). For more on the material presented here, concerning Einstein's time as an undergraduate and in the years leading up to 1905, see Isaacson (2007, pp. 32-89) and Pais (1982, pp. 35–48).
57. Curiously, it seems this 1901 dissertation was lost, and it is not known what the topic was.
58. Einstein's first child was never mentioned in any of his correspondence, save a few letters between Einstein and Maric. She was "discovered" by John Stachel only in 1986. See Isaacson (2007, pp. 72–77).
59. See Galison (2003) for more about Einstein's time at the patent office.
60. For more on Einstein's *annum mirabilis*, see Isaacson (2007, pp. 90–139). The

papers are published together as Einstein (2005), with a valuable introduction by Stachel (2005).

61. Originally published as Einstein (1905a); see also note 60. Einstein's Nobel Prize was for discovering the "law of the photoelectric effect," which is addressed in this paper. However, the photoelectric effect is a minor part of the paper, and the main emphasis is on black-body radiation and light quanta. I am grateful to John Norton for pointing this out to me.
62. This paper was published the following year, as Einstein (1906).
63. The paper was published as Einstein (1905b); see also note 60. Remarkably, Einstein was in many ways anticipated by a little-known French mathematician named Louis Bachelier who, in a dissertation submitted to Sorbonne in 1900, developed a closely related theory of what we now recognize as Brownian motion. Rather than using it to model the behavior of molecules suspended in a fluid, however, Bachelier was aiming to model the statistical properties of government bonds traded at the Paris *Bourse*, for the purposes of developing a method for pricing options on those bonds. See Bachelier (2006 [1900]) and Weatherall (2013).
64. Originally published as Einstein (1905c); see also note 60.
65. Originally published as Einstein (1905d); see also note 60.
66. As with Newton's analysis of space and time in the *Principia*, I find DiSalle's (2006) discussion of Einstein's reanalysis of space and time compelling.
67. At least, this was true in Continental Europe. In the United Kingdom, the reception was more mixed, as the Maxwellians defended the aether. The reception of special relativity is described in Pais (1982, pp. 149–155) and Isaacson (2007, pp. 140–145). For a detailed account of the reception of relativity theory, see the many articles in Glick (1987) and Galison et al. (2001).
68. The classic biography of Maxwell, written by his close friend Lewis Campbell, is Campbell and Garnett (1882); more recent biographies are Everitt (1975), Tolstoy (1981), Goldman (1983), Hendry (1986), and Mahon (2003). Harman (1998) provides a detailed discussion of Maxwell's work. In a recent survey of one hundred prominent physicists, Maxwell was named third greatest, after Einstein and Newton! My personal ranking would round out the top five with Euler and Dirac . . . not that anyone asked.
69. Maxwell's original paper on the electromagnetic field is Maxwell (1865); see also his magnum opus, *A Treatise on Electricity and Magnetism* (Maxwell, 1891). For more information on the history of electricity and magnetism up to and including Maxwell's theory, see Whittaker (1951), Meyer (1972), Heilbron (1979), Darrigol (2000), and Baigrie (2007); briefer overviews are given by Buchwald (2013a; 2013b) and Steinle (2013). Modern textbook presentations of Maxwell's theory are given by Purcell (1985) and Jackson (1999); for more depth, see Hehl and Obukhov (2003).
70. I am grateful to Erik Curiel for pointing this out to me.
71. In addition to the historical works cited in note 69, see Forbes and Mahon

(2014) for the history of Faraday and Maxwell's contributions to unifying electric and magnetic phenomena.

72. My description of Maxwell's understanding of the electromagnetic field comes directly from Maxwell (1865). It is remarkable that Maxwell writes of "so-called vacua" when describing experimentally produced vacua, emphasizing that these regions are not so empty as they appear to be (Maxwell 1865, p. 460). For a fascinating history of the field concept, stretching back to Newton, see Stein (1970); see Berkson (1974) and (especially) Nersessian (1984) for detailed analyses of the evolution of the field concept from Faraday to Einstein. Hesse (1961) offers a grander history of the concept of field, extending back to the ancient Greeks.
73. The classic history of aether theories is given by Whittaker (1951). See also Schaffner (1972) and Darrigol (2000).
74. For a detailed history of the air pump, and its importance for the development of science in the seventeenth century, see Shapin and Schaffer (1985).
75. As remarked in note 72, this inference concerning vacuum chambers comes from Maxwell himself.
76. Later in life, at least, Einstein recalled very early reflections on the motion of light (namely, his famous "riding on a beam of light" thought experiment) as his starting point on the path to relativity theory (Einstein, 1949, p. 49). Of course, Einstein's own recollections of the developments of his ideas, nearly fifty years after the fact, are to be viewed with some suspicion. But there *is* very strong evidence that Einstein arrived at special relativity by reflecting on conceptual problems in electromagnetism, and particularly on the motion of electromagnetic radiation (of which light is just an example). This case is laid out clearly by Norton (2014), who argues in particular that the thought experiment mentioned above, which Einstein recalled later in life, really was significant, as an argument against an emission theory of light. For more detail, see Norton (2004); briefer accounts are given by Isaacson (2007, pp. 113–116), and Pais (1982, pp. 111–134). Other perspectives on the origins of special relativity are given by Holton (1988, pp. 191–236) and Galison (2003).
77. This was established by 1676 by Ole Christensen Rømer. See Romer and Cohen (1940).
78. Important examples of these experiments include the Michelson-Morley experiment (Michelson & Morley, 1887) and the Trouton-Noble experiment (Trouton & Noble, 1904). A nice overview of the influence these had on Einstein is given by Stachel (1987).
79. This is how Einstein begins Part I of the 1905 paper, and it is useful for conceptual and pedagogical reasons. But there is good reason to think that this is *not* how Einstein himself worked his way to special relativity. See Norton (2014).
80. For more on Einstein's reaction to Minkowski's work, see Isaacson (2007, pp. 132–133) and Pais (1982, pp. 151–152). A history of Minkowski's own thinking

on space-time is given by Galison (1979); see also Walter (1999). Minkowski's papers on relativity and related topics have recently been collected as Minkowski (2012). A space-time diagram first appeared as a slide in a lecture Minkowski gave in Cologne in 1908; the lecture was later published in 1909 (Minkowski, 1909) and appears as "Space and Time" in Minkowski (2012).

81. This quote from Einstein, which is frequently repeated, was apparently first published by Pais (1982, p. 152), who claims it was related to him by Bargmann (from memory). It is hard to know what to make of this. But whatever Einstein's initial reaction to Minkowski space-time, he quickly adopted the geometrical methods Minkowski used. I am grateful to John Norton for pointing out the questionable provenance of this remark. The following quote from Minkowski is from Isaacson (2007, p. 132).
82. In my view, the best pedagogical, but technically accurate, description of the geometry of Minkowski space-time is given by Malament (2009).
83. These examples are supposed to conjure up assumptions about distant simultaneity. But they refer to events extended in time, and with relatively short distances between them, and so *parts* of the events will be timelike related. This is precisely why Newtonian notions of simultaneity work so well in ordinary life: the distance light travels in a psychological instant is very large compared to the objects we usually encounter. This point was made very clearly by Stein (1991); see also Callender (2008).
84. This construction for measuring distances can be turned around to provide a way of measuring *duration* (i.e., one can construct a "light clock" by allowing a ray of light to bounce back and forth between mirrors). This construction can be generalized to general relativity as well (Fletcher, 2013).
85. See Einstein (2005, p. 124). Curiously, many of the empirical results we associate with relativity theory had been worked out somewhat earlier, by Hendrik Lorentz (1895; 1899), who did *not* take the aether to be superfluous. (See also FitzGerald [1889] and Poincaré [1906)]. At issue, then, is not whether the aether is compatible with the predictions of relativity theory, but rather whether Einstein's theory or Lorentz's theory provides better explanations of these results. This is laid out nicely by Janssen (2002a; 2002b); see Brown (2005) for a different view on these explanations. Poincaré arguably came even closer to anticipating relativity theory, though his perspective on the aether arguably stood in his way; see Stein ("Physics and Philosophy Meet: the Strange Case of Poincaré," unpublished). Finally, note that while throughout his career, Einstein maintained that there is no *luminiferous* aether of the sort imagined by Maxwell and the Maxwellians, his views on whether there is an aether in some more general sense changed. For instance, in a lecture delivered in 1920, titled, "Ether and Relativity," Einstein argued that in general relativity, space-time has (dynamical) physical properties and thus may be conceived as a kind of physical entity—as Einstein put it, an aether (Einstein, 1922). (See also Einstein [1924].) This idea, that empty space-time in

general relativity has qualitative features that challenge our intuitive idea of "nothingness," will play an important role in what follows. (I am grateful to Michel Janssen for some help with the citations in this note.)

86. For a readable review of current science regarding the Cosmic Microwave Background, including the history of its discovery, see Durrer (2015).
87. My reasons for preferring this second view concern the formulation of the theory in the language of fiber bundles, according to which the electromagnetic field is the curvature of a family of bundles over space-time. From this perspective, vanishing electromagnetic field just means certain bundles have flat connections, and not that they have ceased to exist (in whatever sense one wants to say they "exist" in the first place). This formulation is valuable because it allows one to generalize electromagnetism to describe other fundamental forces, including the weak and strong nuclear forces. For more, see Bleecker (1981), Palais (1981), and Weatherall (2015).
88. Recall note 67. Einstein's trajectory toward an academic position, and his subsequent moves, are described in Isaacson (2007, pp. 149–155, 158–188).
89. See Isaacson (2007, p. 143).
90. See Isaacson (2007, p. 144) and Neffe (2009, p. 155).
91. For a biography of Grossman, with a discussion of his contributions to general relativity, see Sauer (2015). The Einstein-Grossman collaboration is also described in detail by Pais (1982, pp. 208–227), and somewhat more briefly by Isaacson (2007, pp. 192–202). See also the more general references in note 92 concerning the origins of general relativity. Incidentally, Grossman was not Einstein's only colleague at ETH Zurich who would make important contributions to general relativity: Hermann Weyl, whose *Space-Time-Matter* (Weyl, 1922) crystalized the modern geometric understanding of general relativity—and first presented Galilean space-time—was also a recent hire in the mathematics department.
92. The definitive source on the origins of general relativity is the massive four-volume collection of essays and commentary edited by Renn (2007). An excellent, detailed, but still readable overview of Einstein's path to relativity theory, including his many missteps, is given by Janssen (2014); earlier influential treatments were given by Norton (1984) and Stachel (1989). I should emphasize that I am whitewashing the history in my description of the motivations for general relativity. Einstein's development of general relativity was originally motivated by a desire to extend the "special principle of relativity" that he had established in the 1905 papers, according to which any nonaccelerating state of motion is indistinguishable from any other, to a "general principle of relativity" (hence the name), that would accommodate *any* state of motion, accelerating or not. Gravitation was supposed to be the key to making this work, via Einstein's so-called equivalence principle, which states (very roughly) that accelerated motion is, at least locally, indistinguishable from the effects of gravity. As Janssen (2014) nicely argues, Einstein made

many attempts to achieve "general relativity" in this sense, but failed in all of them, and "general relativity" is really a misnomer for the theory that resulted. (I am grateful to John Norton for attempting to keep me honest here.)

93. It seems Einstein had already made a connection to non-Euclidean geometry, but that he did not yet know about Riemannian geometry, which is the key to general relativity. See the references in note 91 for further details on what Einstein knew already and what Grossman contributed. Once again, I am grateful to John Norton for help on this.
94. Definitive treatments of the "modern" geometrical perspective on general relativity are given by Hawking and Ellis (1973), Wald (1984), and Malament (2012); see also Misner, Thorne, and Wheeler (1973). These also provide introductory treatments to the relevant differential geometry. By far the best sustained nontechnical treatment of the geometry of relativity theory is given by Geroch (1978).
95. The strategy pursued here, of developing the physical significance of space-time curvature via the motion of small bodies and light beams, is originally due to Weyl (1922); an alternative perspective, which I find less perspicuous, focuses on rigid bodies and clocks. See also Malament (2012, pp. 119-128). The status of the physical principles underlying this interpretation—namely, the so-called geodesic principle, which states that small bodies follow straightest curves in space-time—has recently been a matter of some dispute: for an overview, see Weatherall (2011).
96. In fact, although the idea that gravitational influences are a manifestation of space-time geometry originated with general relativity, soon after, Cartan (1923; 1924) and Friedrichs (1927) showed that Newtonian gravitation, too, could be understood as a theory in which space-time is curved, and gravitational influences are a manifestation of that curvature. See Malament (2012) for further details, and see Weatherall (2016c).
97. Some physicists and others call Einstein's equation "Einstein's *equations*," with an "s." I prefer not to speak this way: as I understand it, Einstein's equation is a single equation relating several tensors. In a given coordinate system, these tensors have ten independent components, but that is immaterial: for instance, F=ma is a vector equation; we do not speak of three independent equations.
98. Overviews of Einstein's struggles with vacuum solutions, which were tied up with his views on Mach's principle and his attempts to develop a relativistic cosmology, are given by Hoefer (1994), Janssen (2014, pp. 198–208), and Smeenk (2014). Much of what we know about Einstein's views on vacuum solutions during the early history of general relativity comes from the correspondence between Einstein and Willem de Sitter, published in Einstein (1998); see also Schulmann et al. (1998). The realization that specifying the distribution of matter in the universe was not sufficient to determine the

curvature occurred in fits and starts. First there was the Schwarzschild solution, which required additional assumptions about the universe at spatial infinity (Schwarzschild, 1916); then came the de Sitter solution (de Sitter, 1917); and finally, Weyl's general characterization of the part of the curvature of space-time that can be nonzero even when there is no matter present (Weyl, 1922).

99. See Schwarzschild (1916); see also Hawking and Ellis (1973), Wald (1984), and Earman (1995) for details and discussion.
100. The idea that there is a black hole in Schwarzschild space-time was not immediately recognized. Indeed, a proper accounting of singularities in general relativity, including black holes, did not come until the 1960s and 1970s. For a discussion of Einstein's understanding of singularities, see Earman and Eisenstaedt (1999); for the story of how the modern understanding of singularities came about, culminating in the Geroch-Hawking-Penrose singularity theorems, see Earman (1995; 1999); see also Curiel (1999).
101. The solution was published by de Sitter (1917), but only after he presented it in his correspondence with Einstein (Schulmann, et al., 1998). Note that de Sitter space-time is not actually a solution to Einstein's equation; rather, it is a solution to a modified equation with a so-called cosmological constant, which Einstein introduced in 1917 (Smeenk, 2014). From a modern perspective, there is good reason to think that the "cosmological constant" term should really be understood as a contribution to the energy-momentum of space-time, in which case we would not think of de Sitter space-time as a vacuum solution at all. The important point for the current discussion, however, is that *Einstein* seemed to take it as an example of a nontrivial vacuum solution. See also note 98.
102. For more on how Einstein built general relativity by analogy with Maxwell's theory, see Renn and Sauer (2007) and Janssen and Renn (2007).
103. The fact that gravitational waves can *transfer* energy, in the sense of increasing the energy of a material system, does not mean that gravitation waves themselves *carry* energy. In fact, there are serious difficulties associated with attributing any kind of (local) energy or momentum with a gravitational wave. Thank you to Sam Fletcher for keeping me honest here.
104. Recall note 85. For histories of Einstein's attempts to develop a unified field theory, see van Dongen (2010) and Sauer (2014).
105. For more on the early history of gravitational waves, including Einstein's role in developing their theory (and the story told here), see Kennefick (2007); a briefer overview appears as Kennefick (2014).
106. The paper was ultimately published as Einstein and Rosen (1937).
107. A history of the *Physical Review* is given by Hartman (1994); see especially p. 153 for an assessment of the journal's influence by the 1930s.
108. The story of how Einstein left Europe and ended up in Princeton, New Jersey, is told by Isaacson (2007, pp. 394–447).

109. The LIGO results were published as Abbott et al. (2016). For a review of ground-based searches, including LIGO, see Riles (2013); space-based approaches are discussed by Gair et al. (2013).

CHAPTER 3

110. I rely mostly on Farmelo (2009) for biographical material on Dirac throughout the chapter; I also draw on Kragh (1990), Schweber (1994, pp. 11–33, 56–72), and Pais (1998), all of whom give more sophisticated treatments of Dirac's scientific work. See also the recollections collected as Kursunoglu & Wigner, 1990). The circumstances under which Dirac wrote down his eponymous equation are described by Farmelo (2009, pp. 140–148); its lasting importance is nicely captured by Wightman (1972). Dirac described his anxieties about the Dirac equation in a 1963 interview with Thomas Kuhn (American Institute of Physics, 1963), also quoted by Schweber (1994, pp. 59–60); Kragh (1990, p. 61) expresses skepticism about Dirac's recollections on this point. Given how many contributions Dirac made during his career, it is perhaps tendentious to say that the Dirac equation was his *most* significant discovery—though I don't know of any others that are strictly *more* significant. The history of quantum field theory, and Dirac in particular, have received far less scholarly attention than they deserve, and so there is nothing like the rich and extended literature on these topics that one finds on general relativity and Einstein. Schweber (1994) is the principal text, and it is my source for most of the history of science described here. See also Darrigol (1984), Miller (1994), and the first two chapters of Duncan (2012) for further details on the early history of quantum electrodynamics, including Dirac's most important contributions; for more on the history of QED in the late 1940s and into the 1950s, see the early chapters of Kaiser (2005). A brief overview of the history of quantum field theory from Dirac to the Standard Model is given by Kuhlmann (2015).
111. See also van Vleck (1972) for an amusing story of how they discovered his name during his visit to Minnesota.
112. In 1931, Nevill Mott, a Cambridge compatriot who would go on to win a Nobel Prize in physics in 1977, wrote to his parents: "Dirac is rather like one's idea of Gandhi. He is quite indifferent to cold, discomfort, food, etc." (Mott, 1986, p. 42).
113. The interview appeared in the *Wisconsin State Journal*. It is quoted in its entirety by Schweber (1994, pp. 18–20).
114. For a discussion of how Dirac's approach to quantum mechanics related to his contemporaries, see Kragh (1990, pp. 14–47); his time in Copenhagen and Göttingen, and his colleagues' reactions to him, are described by Farmelo (2009, pp. 107–130). Dirac's dissertation was the first on the *new* quantum theory; others—such as Jordan and de Broglie, both in 1924—had written

dissertations on the older theory. As described subsequently in the main text, the history of quantum mechanics is usually traced back to Max Planck's work on black-body radiation (Planck, 1900a; 1900b; 1901); it was then developed by Einstein, Niels Bohr, Arnold Sommerfeld, and many others over the following two and a half decades into what is now known as the "old quantum theory" (ter Haar, 1967). The "new quantum theory," meanwhile, was developed in the period from 1925 to 1927, beginning with the work of Werner Heisenberg, Max Born, and Pascual Jordan (Heisenberg, 1925; Born & Jordan, 1925; Born, Heisenberg & Jordan, 1926) and Erwin Schrödinger (Schrödinger, 1926a; 1926b; 1926c), and culminating in the textbook treatments of Dirac (1930) and von Neumann (1932). See Kuhn (1978) for a detailed history of black-body radiation and early quantum theory and Kragh (2012) for the history of the Bohr model of the atom; a more general overview of the history of the subject is given by Jammer (1966), and an exhaustive treatment is provided by Mehra and Rechenberg (1982–2001).

115. Dirac was born in Bristol, to an English mother, but his father was Swiss and technically Dirac was born a Swiss citizen. He was naturalized in 1919 (Farmelo, 2009, p. 34).
116. See Farmelo (2009, p. 125).
117. For more on Einstein's views on quantum mechanics, see Howard (1985; 1990), Fine (1986), Stachel (1986), and Lehner (2014). See also Bohr (1949) and the references in note 127. The idea that quantum mechanics is "incomplete" is most famously associated with the so-called EPR paper (Einstein, Podolsky, & Rosen, 1935).
118. There are many ways of presenting the basic structure of quantum mechanics. The presentation here is in the tradition of von Neumann (1932), and most closely follows Jauch (1968) and Stein (1972)—albeit without the mathematics. Gentle, self-contained introductions with different emphases, using a minimal amount of mathematics, are given by Albert (1992), Susskind and Friedman (2014), and Bub (2016); a more sophisticated, but still conceptually oriented, introduction is given by Hughes (1989).
119. See Kent (2001) and Stanton (2015).
120. Or, as philosophers would put it, the probabilities associated with a statistical description of a classical state are *epistemic*, rather than objective. For more on interpretations of probability, see Hájek (2012) and Childers (2013). The claim in the text that follows, that quantum probabilities are not epistemic in this sense, is interpretation-dependent: for instance, in Bohm's theory, quantum probabilities *are* epistemic. (See note 121 for more on interpretations of quantum theory.)
121. A great deal has been written on the measurement problem, by physicists and philosophers alike. Some classics are Shimony (1963), Stein (1972), and the various essays on quantum mechanics by John Bell, stretching back to the 1960s and published together as Bell (2004). See also Jammer (1974), Fine

(1986), Redhead (1987), Hughes (1989), van Fraassen (1991), Albert (1992), Maudlin (1994), Bub (1997), Barrett (1999), and Wallace (2012) for discussions of different perspectives—and defenses of various solutions to the problem. For examples of interpretations in which one changes the rules of probability, see Fine (1982a; 1982b); see also Feintzeig (2014). Alternative theories that are empirically similar or equivalent include Bohm's theory (see Bohm [1952]; Maudlin [1994]; and Dürr and Teufel [2009]) and the GRW theory (see Ghirardi, Rimini, and Weber [1986]); the most important no-collapse theory is that of Everett (see Everett [1957]; Barrett [1999]; and Wallace [2012]).

122. Or at least, they were widely reported to have provided such evidence, in part because of Eddington's influence. But there is a long-standing controversy concerning just what was shown by the eclipse data—in part because Eddington himself admitted to being "not altogether unbiased" (Eddington, 1920, p. 116). Again, for details on the 1919 expedition and its significance, including the controversy, see Earman and Glymour (1980), Stanley (2003), and Kennefick (2012). For biographies of Eddington, see Douglas (1956), Chandrasekhar (1983), and Stanley (2007). For more on the climate in mathematical physics at Cambridge during this period, especially regarding Eddington and the reception of relativity theory, see chapter 9 of Warwick (2003).
123. The book on relativity was published as Dirac (1975). The first paper on Dirac's quantum theory of the electromagnetic field was Dirac (1927), though as Dirac observes in the paper, his treatment is not entirely relativistic, in part because the electrons interacting with the electromagnetic fields are treated as nonrelativistic particles; this problem is addressed in Dirac (1928), where the Dirac equation is introduced. See Schweber (1994, pp. 23–32) and Duncan (2012, pp. 31–37) for discussion of Dirac (1927).
124. The best treatment of QED for a general audience is surely Feynman (1985a). For textbook treatments of QED, see the classics by Jauch and Rohrlich (1955) and Bjorken and Drell (1964), as well as the more recent book by Milonni (1994); QED is also treated in depth in more general books on quantum field theory, such as Peskin and Schroeder (1995), Weinberg (1995–2000), Duncan (2012), and Schwartz (2013).
125. For more on Einstein in the 1920s, see Isaacson (2007, pp. 309–383); references on his views concerning quantum physics are given in note 117.
126. For more on the 1927 Solvay conference, including a full translated transcript, see Bacciagaluppi and Valentini (2009), as well as the references in note 127.
127. The disagreements between Einstein and Bohr, both of whom had made major contributions to the old quantum theory (see note 114), were especially pronounced at the 1927 Solvay conference, but in a sense this was only the beginning. Einstein and Bohr continued to debate, in person and in published work, until the end of their lives—or at least until 1948, when Bohr came to visit the Institute for Advanced Study in Princeton. During his stay, he wrote down his recollections of the Solvay debate (Bohr, 1949). For more

on the dispute between Einstein and Bohr, see Hooker (1972), Jammer (1974), Pais (1982, pp. 440–459), Fine (1986), Beller and Fine (1994), Whitaker (1996), Beller (1999), Landsman (2006), Bacciagaluppi and Valentini (2009), and Lehner (2014). For more on Einstein's and Bohr's philosophical views more generally, see (for Bohr) Folse (1985), Murdoch (1989), Faye (1991), and Faye and Folse (1994), and see Howard (2014) and references therein (for Einstein). For biographies of Bohr, see Silverberg (1965) and Pais (1993).

128. That Dirac was bored with the debate does not mean he did not take sides. In fact, somewhat surprisingly, it seems Dirac sided with Einstein (Bokulich, 2008).
129. Einstein apparently uttered and wrote variants on this quote many times, though the standard reference is to a letter to Max Born on December 4, 1926; see also Bohr (1949, p. 218) and Isaacson (2007, p. 326) and the associated note.
130. See Heisenberg (1971, pp. 85–86). The full quote is improbably lengthy and precise, given that Heisenberg was recalling it many years later. Still, one might hope Heisenberg accurately captured Dirac's view, even if the words were not really his. Pauli's rejoinder is quoted in Schweber (1994, p. 15).
131. For a recent discussion of how Einstein drew on the work of others, see Janssen and Renn (2015).
132. See the discussion and references in note 114.
133. Einstein's work on light quanta was published as 1905a. For more on Einstein's contributions to quantum mechanics, see Pauli (1949), Kuhn (1978), Stone (2013), and Darrigol (2014). The view presented here, that quantum field theory should be traced all the way back to Planck and Einstein's work on radiation, is convincingly defended by Duncan (2012).
134. For more on Jordan's life and contributions, see Schweber (1994, pp. 5–11, 33–39) and Wise (2003); see also Darrigol (1986), Duncan and Janssen (2008; 2009; 2012), and chapter 2 of Duncan (2012). Despite his important contributions, Jordan has received considerably less attention than the other inventors of quantum theory, for a number of reasons. Perhaps the most important is that he had a breakdown in 1930 and largely left physics, turning his attention to more speculative matters in biology, psychology, and other fields. Later in the 1930s, he joined the Nazi Party, which also surely contributed to his assessment by later physicists.
135. This paper was Born, Heisenberg, and Jordan (1926).
136. For more on the idea of "wave-particle dualism," see Wheaton (1983) and Büttner et al. (2003); for more on Schrödinger and the history of wave mechanics, including the origins of the theory in the work of de Broglie, see volume 5 of Mehra and Rechenberg (1982–2001), Darrigol (2013), and Renn (2013). See also the references to Schrödinger's work in note 114.
137. The so-called Born rule for deriving probabilities from solutions to Schrödinger's equation was first proposed by Born (1926). For more on the

debates concerning the Born rule, see Jammer (1966) and Beller (1999), as well as (more generally) the references on the measurement problem in note 121; for a discussion of Schrödinger's own views, see Bitboll (1996).

138. See the references in note 134. Jordan developed the theory of quantized matter waves in a number of papers, alone and in collaboration, beginning with Jordan (1927).
139. The details of how Dirac proceeded are given by Schweber (1994, pp. 56–60). See also Kragh (1981; 1990, pp. 57–62) and Moyer (1981a). Duncan and Janssen (2009) provide further details on the transformation theory that motivated Dirac's calculations, which Dirac and Jordan developed independently. The history here is quite complex. As I note in the text, it was clear before Jordan made his proposal about quantizing matter waves that a relativistic analogue to the Schrödinger equation was necessary, and Dirac had considered this problem as early as 1926, before writing the paper on QED. Moreover, even in the QED paper, he performed calculations with the Schrödinger equation that were formally suggestive of Jordan's proposal, albeit with a different interpretation. So it is not right to say that Dirac was trying to accomplish a goal that Jordan had set, or that he was working in response to anything Jordan had done—he would have sought the Dirac equation one way or the other. Indeed, at least initially, Dirac opposed Jordan's proposal. Rather, Jordan's work in 1927 was a crucial contribution to interpreting the Dirac equation once it was produced, as well as an important step forward in its own right in developing the conceptual framework of quantum field theory. I am grateful to Tony Duncan for help on the relationship between Dirac's and Jordan's thought during this period.
140. The Klein-Gordon equation was originally discovered by Schrödinger, but never published; it was then discovered independently, and published, by Oskar Klein (1926), with early further developments by Walter Gordon (1926) and Vladimir Fock (1926). See Kragh (1990, pp. 49–57) and Schweber (1994, p. 57) for more on the history.
141. See Kragh (1981; 1990, pp. 54–56) and Schweber (1994, pp. 57–58) for discussions of these reasons; Dirac discusses them himself in the interview with Kuhn (American Institute of Physics, 1963). It seems that the main reason was that Dirac was committed to what he called the "transformation theory," which he believed was incompatible with the Klein-Gordon equation.
142. The paper was published as Dirac (1928). The published version records the date of receipt as January 2, 1928, but Farmelo claims it was submitted on New Year's Day.
143. These calculations showed that the Dirac equation could account for the spectral lines of hydrogen; the calculations were done by Charles Darwin and Walter Gordon, Dirac's friend in Cambridge and one of the originators of the Klein-Gordon equation, respectively. The reception of the Dirac equation is described by Moyer (1981b), Kragh (1990, pp. 62–65), Schweber (1994, p. 58),

and Farmelo (2009, pp. 143–146). Again, see Wightman (1972) for an account of why the Dirac equation was so important.

144. The vicissitudes of Dirac's "hole theory," leading up to his successful reinterpretation of the negative energy states and prediction of positrons, are described in detail by Moyer (1981b; 1981c), Kragh (1990, pp. 87–117), and Schweber (1994, pp. 56–69). Saunders (1991) provides a more conceptual take on the hole theory and its lasting influence; see also Pashby (2012) for a discussion of the broader philosophical import of the discovery of the positron.
145. The quote is from Dirac (1931, p. 65); see also Schweber (1994, p. 66).
146. The discovery of the positron is detailed by Hanson (1961), de Maria and Russo (1985), and Roque (1997). See also Leone and Robotti (2010) for further details on the discovery of the positron, including an earlier near-miss.
147. The Anderson papers are Anderson (1932a; 1932b); the other physicists who observed the positron soon after Anderson were Blackett and Occialini (1933), who were both based in Cambridge and had the benefit of discussing their work with Dirac. Several authors, most notably Kuhn (1962), have argued that it was Blackett and Occialini who truly discovered the positron, because it was only the latter authors who truly saw it *as* the positron, even though Anderson observed it first. See Roque (1997) for a discussion.
148. The development of quantum field theory during the 1930s is described in chapter 2 of Schweber (1994).
149. For biographical material on both Feynman and Schwinger, including detailed analyses of their work individually and comparatively, see chapters 3–5, 7, and 8 of Schweber (1994). For more on Feynman's life and work, see the scientific biography by Mehra (1994); see also Gleick (1992) and Krauss (2011), as well as the recollections compiled by Brown and Rigden (1993). Feynman (1985b; 1988) includes some humorous and insightful autobiographical notes. For more on Schwinger, see the scientific biography by Mehra and Milton (2000).
150. At Schwinger's sixtieth birthday party, Feynman described how the two of them would talk regularly, comparing results even though they didn't understand one another's methods. As Feynman put it then, "Many people joked we were competitors—but I don't remember feeling that way" (Schweber, 1994, p. 445). For evidence that he *did* feel that way, at least sometimes, see Schweber (1994, p. 427). For Schwinger's side, meanwhile, see, for instance, Schweber (1994, pp. 352–355).
151. This Gell-Mann quote is from Johnson (2000), in reference to *Surely You're Joking, Mr. Feynman* (Feynman, 1985b).
152. See Mehra and Milton (2000, pp. 548–554).
153. Quoted in Schweber (1994, p. 445).
154. The claim that Dirac was, secretly, a geometric thinker is based on a fascinating paper by Galison (2000); the claim that Einstein was an algebraic thinker, meanwhile, comes from Norton (2007; and conversation).
155. Feynman diagrams are explained in any modern textbook on quantum field

theory; see the references in note 124. For a discussion of the history of Feynman diagrams, see Kaiser (2005) and Wüthrich (2010). The difference between how Feynman and Schwinger thought is evident from reading their papers, but it is perhaps worth noting that Schweber (1994, p. 471) makes the same classification. The claim that Schwinger discouraged his students from using Feynman diagrams is from Kaiser (2005, pp. 104–108).

156. The Wheeler-Feynman theory is described by Schweber (1994, pp. 380–389); see also Wheeler and Feynman (1945; 1949).
157. For Schwinger's views on particles and fields, see, for instance, Schwinger (1969, pp. 24-36); see also Schweber (1994, pp. 355–365).
158. There are long-standing debates in physics and philosophy of physics concerning the status of particles, fields, and "quanta," or particlelike field excitations, in quantum field theory. For instance, see Hegerfeldt (1974), Birrell and Davies (1982, pp. 48–59), Weinberg (1987), Wald (1994, pp. 46–52), Huggett (1994), Teller (1995), Malament (1996), Clifton and Halvorson (2001), Halvorson and Clifton (2002), Fraser (2008), and Baker (2009). Lupher (2010) provides a nice overview of the issues. Puzzles about the particle and field interpretations of quantum field theory go back to the very origins of the theory and relate to earlier issues concerning wave/particle duality. See, for instance, Duncan and Janssen (2008).
159. Detailed discussions of the vacuum in quantum field theory, with a range of interpretations, are given by Redhead (1983; 1988; 1994; 1995), Saunders (1991; 2002), Sciama (1991), Teller (1993; 1995), Rugh, Zinkernagel, and Cao (1999), Summers (2011); and Boi (2011); of course, see, too, the textbooks on quantum field theory cited in note 124. Insofar as I take sides in the discussion above—which is really meant to be expository—the view I present is probably closest to Redhead. Note that the interpretation of the vacuum is closely related to other interpretational issues in quantum field theory, such as the possibility of a "particle interpretation" of the theory and the status of "quanta" and "field excitations." See note 158.
160. In fact, there is an important subtlety here: although there *is* an observable that we might think of as a "global" number observable, there do *not* exist local number observables, (i.e., ones associated with a particular region of space and time, as opposed to the whole universe for all eternity). This is a consequence of a result known as the Reeh-Schlieder theorem, which implies that there is no way to define local observables that always yields "0" for the number of particles in the vacuum state. (See Redhead [1995] and Summers [2011] for details.) On the other hand, when we build particle detectors, we are measuring *something*, and we are doing it within some local region of space and time. Whatever that something is must be represented by some local observable in the theory. It turns out that these observables can be chosen to "approximate" null results arbitrarily well in the vacuum state. (See chapter 6 of Haag [1992]; section 4 of Buchholz [2000]; and the final section of

Halvorson and Clifton [2002] for details on this sense of "approximate" local number observable; more recently, see Arageorgis and Stergiou [2013] and Valente [2015].) When I write of local number observables above, and of what one would expect from measurements with particle detectors, I am assuming that the measurements are represented by local observables that are "approximate" number observables in this sense. The Reeh-Schlieder theorem, then, is what supports my claim that these measurements are never guaranteed to yield "0."

161. Tryon (1973) famously argues that the universe *is* a vacuum fluctuation; Albert (1989) provides a curious argument to the effect that we could never know if the universe is described by the vacuum state. See also the discussion in note 160. All that said, on standard views, the probability of finding the world as we see it in the vacuum state is vanishingly small.
162. There are others—for instance, the Lamb shift and the Casimir effect. These are described in detail in chapter 2 of Milonni (1994). The early history of vacuum polarization is described by Schweber (1994, pp. 86–87).
163. For more on the Shelter Island conference, see Schweber (1994, pp. 156–205, 303–318, 411–414); Bethe's calculation of the Lamb shift, which is really a result of the modification to the Coulomb potential from charge renormalization, is described by Schweber (1994, pp. 228–231). The connection between vacuum polarization and the Lamb shift is explained clearly by Schwartz (2013, pp. 300–314); see also Milonni (1994, pp. 314–315).
164. Brillouin (1960) remains the standard source for wave propagation in a dielectric medium.
165. For instance, see Odom et al. (2007).
166. Wightman (1989) provides a personal history of the axiomatic approach to quantum field theory during the 1950s, and Fredenhagen et al. (2007) gives a nice overview of the current state of affairs; see also Streater (1988) for motivation for axiomatizing quantum field theory, and Streater and Wightman (2000) for more detail on what was accomplished in the early years of the axiomatic program. Haag (1996) and Araki (1999) provide the classic accounts of algebraic methods in quantum field theory, and Brunetti et al. (2015) provides an up-to-date account. See Glimm and Jaffe (1987) for the classic in-depth discussion of so-called constructive quantum field theory, where one attempts to produce rigorous mathematical models for the various axiom systems described above; Summers (2012) provides a more recent perspective. More recently, there have been a number of attempts, using somewhat different methods, to put quantum field theory on a rigorous footing, by attempting to more closely mimic the methods of working physicists. For instance, see Hollands (2009), Costello (2011), and Rejzner (2016), though these discussions are far from exhaustive.
167. For instance, see Schweber (1994, pp. 436–445).
168. For more on Wilsonian renormalization, see the influential review article by

Wilson and Kogut (1974) or any of the textbooks in note 124; for more on the "effective field theory" picture that has emerged from Wilson's ideas, see Polchinski (1984; 1992), Georgi (1993), and Cao and Schweber (1993). For many physicists, Wilsonian methods solve all of the truly troubling problems; for general expressions of this attitude, see Wallace (2011) and Duncan (2012). A response is given by Fraser (2011). When I say in the main text that Wilson has not solved all of the problems, what I have in mind is that there are no known (nonperturbative) four-dimensional models of quantum field theories with interactions, and that the perturbation series widely used in physics are not known to converge (and in some cases, are known not to converge). Finding such a model is a so-called Millennium Problem, the solving of which would result in an award of a million dollars; the details of the Millennium Problem were described by Arthur Jaffe, mentioned above, and Edward Witten, the most influential figure in string theory.

169. See Kragh (1990, pp. 183–188) for a discussion of Dirac's reaction to QED in the early 1950s, and also his attempts to develop an alternative theory based on "sensible mathematics." The quotes are taken from p. 184. It is important to emphasize, though, that by "sensible mathematics," Dirac did not necessarily mean "rigorous mathematics." He was much more concerned with calculations that he saw as ad hoc or arbitrary than he was with calculations that were not completely well-defined mathematically. It is not clear how Dirac would have reacted if he had come to appreciate the new attitudes toward renormalization developed during the 1970s. (See note 168.) See chapter 14 of Kragh (1990) for a discussion of Dirac's views on rigor and mathematical beauty. Note, too, that there is a curious parallel between Dirac's later work on QED and Schwinger's: in both cases, they felt that there was something inadequate about the standard theory, and they sought to develop something new that they thought was more principled. And in both cases, the mainstream physics community that had once celebrated them largely ignored the new work because it seemed to merely recapitulate known results.

170. See Summers (2011).

EPILOGUE

171. The events I have in mind are (1) the discovery, in 1973, that a wide class of theories, including quantum chromodynamics, has a property known as "asymptotic freedom," by Politzer (1973) and Gross and Wilczek (1973); and (2) the discovery, at the Stanford Linear Accelerator and at Brookhaven National Laboratory, of the J/psi meson. The first discovery showed that QCD is well-behaved in ways that QED and other quantum field theories are not, making it a particularly attractive theory; the second discovery provided evidence for a new kind of quark—the charm quark—and led to widespread acceptance of the quark theory, and QCD in particular. For more on the history

of the standard model, see the articles in the collection by Hoddeson et al., (1997). For more on the history of QCD in particular, see Cao (2010); Pickering (1984), though a classic, has a rather different perspective from the one adopted here. For textbook presentations of the Standard Model, including QCD, see Cottingham and Greenwood (2007) and Schwartz (2013), as well as the classics by Peskin and Schroeder (1995) and Weinberg (1995–2000).

172. I have in mind Wilczek, Gross, and Politzer, who shared the 2004 prize for the discovery of asymptotic freedom, and Richter and Ting, who shared the 1976 prize for discovering the J/psi meson. See note 171.

173. The history of string theory I tell here relies mostly on Galison (1995); see also Rickles (2014) and the various essays in the collection by Cappelli et al. (2012). Standard textbook introductions to string theory are given by Green et al. (1987) and Polchinski (1998); accounts for a more general audience are given by Greene (1999) and Gubser (2010).

174. For more on these three forces, see the references in note 171.

175. There are a number of strategies for constructing a quantum theory of gravity that have attracted significant attention. For general discussions of the problem of quantizing gravity, with attention paid to all of the major efforts, see Kiefer (2012); Smolin (2001) offers a gentler introduction to the issues. A brief, general history of the subject is given by Rovelli (2001). The currently dominant tradition, stretching back to the 1970s, is string theory; see note 173. The main competitor is known as loop quantum gravity, which originated in the 1980s with the work of Amitabha Sen (1982) and (especially) Abhay Ashtekar (1986). Loop quantum gravity emerged as an outgrowth of the canonical quantization approach of Dirac (1958), Peter Bergman (1966), Bryce DeWitt (1967), and John Wheeler. Textbook presentations of loop quantum gravity are given by Rovelli (2004) and Gambini and Pullin (2011).

176. The problem I am describing here is closely related to the nonexistence of "local" observables in canonical quantum gravity and to the so-called problem of time. See Isham (1992) and Kiefer (2012) for further discussion.

177. Indeed, the history of quantum gravity arguably goes back to the very beginnings of general relativity. A discussion of the early history of the subject is given by Stachel (1999). See Rovelli (2001) for a discussion of the history of canonical quantum gravity, and particularly recognition of the others who contributed to it. (See also note 175.) It is perhaps worth noting that the Wheeler-DeWitt equation first appeared in a paper by DeWitt (1967), where he attributed it to Wheeler and called it the "Einstein-Schrödinger Equation"; in later work, Wheeler always referred to it as the DeWitt equation. While I say in the text that canonical quantum gravity ran out of steam, in many ways loop quantum gravity—which is still an active area of research—is its heir, and so the tradition continues.

178. See Davies and Brown (1992, p. 194). This volume is from a collection of interviews with famous physicists evaluating string theory. It includes a number

of influential defenders—such as Edward Witten, Michael Green, and John Schwarz—as well as some important critics, including both Feynman and Sheldon Glashow, another Nobel laureate theorist who made major contributions to the Standard Model. The most important criticism of string theory—which has plagued the theory since its inception—is that it has not succeeded in making connections with experimental science—especially within the areas of high-energy physics and cosmology that it was intended to address. Particularly strong critiques of string theory were given by Woit (2006) and Smolin (2006), who also complained that string theorists have overhyped the theory and (allegedly) suppressed other research programs. (See also Curiel [2001].) Reaction to these books from string theorists was extremely negative, leading to a public debate sometimes known as the "string wars." Recently, more sober attempts to explain to a lay audience what is compelling about string theory have appeared: for instance, Dawid (2014) and Conlon (2015).

179. Recall notes 125 and 127.
180. See Galison (1995). One can still see this idea of "inevitability" in the interviews with Schwarz and Green in the collection by Davies and Brown (1992); see also Greene (1999).
181. This paper was by Lerche, Lüst, and Schellekens (1987).
182. For instance, the 10^{500} estimate appears in Blumenhagen et al. (2005); Susskind (2003) put the estimate at somewhere between 10^{100} (a "googol") and $10^{10\wedge100}$ (a "googolplex").
183. The idea was presented by Susskind (2003), though it has some prehistory—see, for instance, Schellekens (2006).

Bibliography

Abbott, B. P., et al. (2016). Observation of Gravitational Waves from a Binary Black Hole Merger. *Physical Review Letters, 116*(6), 061102.

Albert, D. (1992). *Quantum Mechanics and Experience.* Boston, MA: Harvard University Press.

Albert, D. Z. (1989). On the Possibility That the Present Quantum State of the Universe Is the Vacuum. In A. Fine & J. Leplin, *PSA 1988: Proceedings of the 1988 Biennial Meeting of the Philosophy of Science Association* (pp. 127–133). Chicago, IL: University of Chicago Press.

Albert, D. Z. (2012, March 23). On the Origin of Everything. *The New York Times*, BR20.

American Institute of Physics. (1963). *Interview of P. A. M. Dirac by Thomas S. Kuhn and Eugene Wigner on 1962 April 1.* Retrieved from American Institute of Physics: https://www.aip.org/history-programs/niels-bohr-library/oral-histories/4575-1

Anderson, C. D. (1932a). The Apparent Existence of Easily Deflectable Positives. *Science, 76*(1967), 238–239.

Anderson, C. D. (1932b). Energies of Cosmic-Ray Particles. *Physical Review, 41*(4), 405–412.

Antognazza, M. R. (2009). *Leibniz: An Intellectual Biography.* Cambridge, UK: Cambridge University Press.

Arageorgis, A., & Stergiou, C. (2013). On Particle Phenomenology without Particle Ontology: How Much Local Is Almost Local? *Foundations of Physics, 43*(8), 969–977.

Araki, H. (1999). *Mathematical Theory of Quantum Fields.* (U. Carow-Watamura, Trans.). Oxford, UK: Oxford University Press.

Ariew, R. (1995). G. W. Leibniz, Life and Works. In N. Jolley, *The Cambridge Companion to Leibniz* (pp. 18–42). Cambridge, UK: Cambridge University Press.

Ariew, R. (2000). Introduction. In G. W. Leibniz, S. Clarke, & R. Ariew, *Correspondence* (pp. vii–xv). Indianapolis, IN: Hackett Publishing Company.

Aristotle. (1984). Physics. In J. Barnes, *The Complete Works of Aristotle: The Revised Oxford Translation* (R. P. Hardie & R. K. Gaye, Trans., Vol. 1, pp. 315–446). Princeton, NJ: Princeton University Press.

Ashtekar, A. (1986). New Variables for Classical and Quantum Gravity. *Physical Review Letters, 57*(18), 2244–2247.

Bacciagaluppi, G., & Valentini, A. (2009). *Quantum Theory at the Crossroads: Reconsidering the 1927 Solvay Conference.* Cambridge, UK: Cambridge University Press.

Bachelier, L. (2006 [1900]). *Louis Bachelier's Theory of Speculation: The Origins of Modern Finance.* (M. Davis & A. Etheridge, Eds.). Princeton, NJ: Princeton University Press.

Baigrie, B. S. (2007). *Electricity and Magnetism: A Historical Perspective.* Westport, CT: Greenwood Press.

Baker, D. J. (2009). Against Field Interpretations of Quantum Field Theory. *The British Journal for Philosophy of Science, 60*(3), 585–609.

Barbour, J. B., & Bertotti, B. (1982). Mach's Principle and the Structure of Dynamical Theories. *Proceedings of the Royal Society of London A, 382*, 295–306.

Barrett, J. A. (1999). *The Quantum Mechanics of Minds and Worlds.* Oxford, UK: Oxford University Press.

Barrow, J. D. (2001). *The Book of Nothing: Vacuums, Voids, and the Latest Ideas about the Origins of the Universe.* New York, NY: Pantheon Books.

Bell, J. S. (2004). *Speakable and Unspeakable in Quantum Mechanics* (2nd ed.). Cambridge, UK: Cambridge University Press.

Beller, M. (1999). *Quantum Dialogue: The Making of a Revolution.* Chicago, IL: University of Chicago Press.

Beller, M., & Fine, A. (1994). Bohr's Response to EPR. In J. Faye & H. J. Folse, *Niels Bohr and Contemporary Philosophy* (pp. 1–31). Dordrecht, Netherlands: Kluwer.

Bergmann, P. G. (1966). Hamilton-Jacobi and Schrödinger Theory in Theories with First-Class Hamiltonian Constraints. *Physical Review, 144*(4), 1078–1080.

Berkson, W. (1974). *Fields of Force: The Development of a World View from Faraday to Einstein.* New York, NY: John Wiley & Sons.

Bernstein, J. (1973). *Einstein.* New York, NY: Viking Press.

Berryman, S. (2011). *Ancient Atomism.* (E. N. Zalta, Ed.). Retrieved from the Stanford Encyclopedia of Philosophy: http://plato.stanford.edu/archives/win2011/entries/atomism-ancient/

Birrell, N. D., & Davies, P. C. (1982). *Quantum Fields in Curved Space.* Cambridge, UK: Cambridge University Press.

Bitboll, M. (1996). *Schrödinger's Philosophy of Quantum Mechanics.* Dordrecht, Netherlands: Kluwer.

Bjorken, J. D., & Drell, S. D. (1964). *Relativistic Quantum Mechanics.* New York, NY: McGraw Hill.

Blackett, P. M., & Occhialini, G. P. (1933). Some Photographs of the Tracks of Penetrating Radiation. *Proceedings of the Royal Society of London A, 139*(839), 699–720.

Bleecker, D. (1981). *Gauge Theory and Variational Principles.* Reading, MA: Addison-Wesley Publishing.

Blumenhagen, R., Gmeiner, F., Honecker, G., Lüst, D., & Weigand, T. (2005). The Statistics of Supersymmetric D-brane Models. *Nuclear Physics B, 713*, 83–135.

Bohm, D. (1952). A Suggested Interpretation of the Quantum Theory in Terms of "Hidden" Variables, I and II. *Physical Review, 85*(2), 166–193.

Bohr, N. (1949). Discussion with Einstein on Epistemological Problems in Atomic Physics. In P. A. Schilpp, *Albert Einstein: Philosopher-Scientist* (pp. 199–242). La Salle, IL: Open Court Press.

Boi, L. (2011). *The Quantum Vacuum.* Baltimore, MD: Johns Hopkins University Press.

Bokulich, A. (2008). Paul Dirac and the Einstein-Bohr Debate. *Perspectives on Science, 16*(1), 103–114.

Born, M. (1926). Zur Quantenmechanik der Stossvorgänge. *Zeitschrift für Physik, 37*(12), 863–867.

Born, M., Heisenberg, W., & Jordan, P. (1926). Zur Quantenmechanik. II. *Zeitschrift für Physik, 35*(8), 557–615.

Born, M., & Jordan, P. (1925). Zur Quantenmechanik. *Zeitschrift für Physik, 34*(1), 858–888.

Boyer, C. B. (1959). *The History of the Calculus and Its Conceptual Development.* New York, NY: Dover.

Brillouin, L. (1960). *Wave Propagation and Group Velocity.* New York, NY: Academic Press.

Brown, H. (2005). *Physical Relativity.* Oxford, UK: Oxford University Press.

Brown, L. M., & Rigden, J. S. (1993). *"Most of the Good Stuff": Memories of Richard Feynman.* New York, NY: Springer-Verlag.

Brunetti, R., Dappiaggi, C., Fredenhagen, K., & Yngvason, J. (2015). *Advances in Algebraic Quantum Field Theory.* Berlin, Germany: Springer-Verlag.

Bub, J. (1997). *Interpreting the Quantum World.* Cambridge, UK: Cambridge University Press.

Bub, J. (2016). *Bananaworld: Quantum Mechanics for Primates.* Oxford, UK: Oxford University Press.

Buchholz, D. (2000). *Algebraic Quantum Field Theory: A Status Report.* Retrieved from arXiv:math-ph/0011044

Buchwald, J. Z. (2013a). Electricity and Magnetism to Volta. In J. Z. Buchwald & R. Fox, *The Oxford Handbook of the History of Physics* (pp. 432–442). Oxford, UK: Oxford University Press.

Buchwald, J. Z. (2013b). Electrodynamics from Thomson and Maxwell to Hertz. In J. Z. Buchwald & R. Fox, *The Oxford Handbook of the History of Physics* (pp. 571–583). Oxford, UK: Oxford University Press.

Buchwald, J. Z., & Feingold, M. (2012). *Newton and the Origin of Civilization*. Princeton, NJ: Princeton University Press.

Büttner, J., Renn, J., & Schemmel, M. (2003). Exploring the Limits of Classical Physics: Planck, Einstein, and the Structure of a Scientific Revolution. *Studies in History and Philosophy of Modern Physics, 34*(1), 37–59.

Callender, C. (2008). The Common Now. *Philosophical Issues, 18*(1), 339–361.

Campbell, L., & Garnett, W. (1882). *The Life of James Clerk Maxwell*. London, UK: Macmillan and Co.

Cao, T. Y. (2010). *From Current Algebra to Quantum Chromodynamics: A Case for Structural Realism*. Cambridge, UK: Cambridge University Press.

Cao, T. Y., & Schweber, S. S. (1993). The Conceptual Foundations and the Philosophical Aspects of Renormalization Theory. *Synthese, 97*(1), 33–108.

Cappelli, A., Castellani, E., Colomo, F., & di Vecchia, P. (2012). *The Birth of String Theory*. Cambridge, UK: Cambridge University Press.

Cartan, É. (1923). Sur les variétés à connexion affine, et la théorie de la relativité (première partie). *Annales Scientifiques de l'École Normale Supérieure, 40*, 325–412.

Cartan, É. (1924). Sur les variétés à connexion affine, et la théorie de la relativité (suite). *Annales scientifiques de l'École Normale Supérieure, 41*, 1–25.

Chalmers, A. (2014). *Atomism from the 17th to the 20th Century*. (E. N. Zalta, Ed.). Retrieved from the Stanford Encyclopedia of Philosophy: http://plato.stanford.edu/archives/win2014/entries/atomism-modern/

Chandrasekhar, S. (1983). *Eddington: The Most Distinguished Astrophysicist of His Time*. Cambridge, UK: Cambridge University Press.

Childers, T. (2013). *Philosophy and Probability*. Oxford, UK: Oxford University Press.

Clifton, R., & Halvorson, H. (2001). Are Rindler Quanta Real? Inequivalent Particle Concepts in Quantum Field Theory. *The British Journal for the Philosophy of Science, 52*(3), 417–470.

Cohen, I. B. (1983). *The Newtonian Revolution*. Cambridge, UK: Cambridge University Press.

Cohen, I. B. (1999). A Guide to Newton's Principia. In I. Newton, I. B. Cohen, & A. Whitman, *The Principia: Mathematical Principles of Natural Philosophy* (pp. 1–370). Berkeley: University of California Press.

Cohen, I. B., & Koyré, A. (1962). Newton and the Leibniz-Clarke Correspondence. *Archives internationales d'histoire des science, 15*, 63–126.

Cohen, I. B., & Westfall, R. S. (1995). *Newton*. New York, NY: Norton.

Conlon, J. (2015). *Why String Theory?* London, UK: CRC Press.

Costello, K. (2011). *Renormalization and Effective Field Theory*. Providence, RI: American Mathematical Society.

Cottingham, W. N., & Greenwood, D. A. (2007). *An Introduction to the Standard Model of Particle Physics* (2nd ed.). Cambridge, UK: Cambridge University Press.

Curiel, E. (1999). The Analysis of Singular Spacetimes. *Philosophy of Science, 66*(Proceedings), S119–S145.

Curiel, E. (2001). Against the Excesses of Quantum Gravity: A Plea for Modesty. *Philosophy of Science, 68*(Proceedings), S424–S441.

Darrigol, O. (1984). La genèse du concept de champ quantique. *Annales de physique*, 433–501.

Darrigol, O. (1986). The Origin of Quantized Matter Waves. *Historical Studies in the Physical and Biological Sciences, 16*(2), 197–253.

Darrigol, O. (2000). *Electrodynamics from Ampère to Einstein*. Oxford, UK: Oxford University Press.

Darrigol, O. (2013). A Few Reasons Why Louis de Broglie Discovered Broglie's Waves and Yet Did Not Discover Schrödinger's Equation. In W. L. Reiter & J. Yngvason, *Erwin Schrödinger—50 Years After* (pp. 165–174). Zürich, Switzerland: EMS Publishing House.

Darrigol, O. (2014). The Quantum Enigma. In M. Janssen & C. Lehner, *The Cambridge Companion to Einstein* (pp. 117–142). Cambridge, UK: Cambridge University Press.

Davies, P. C., & Brown, J. (1992). *Superstrings: A Theory of Everything?* Cambridge, UK: Cambridge University Press.

Dawid, R. (2014). *String Theory and the Scientific Method*. Cambridge, UK: Cambridge University Press.

de Maria, M., & Russo, A. (1985). The Discovery of the Positron. *Rivista di Storia della Scienza, 2*, 237–286.

Descartes, R. (1985). Principles of Philosophy. In J. Cottingham, R. Stoothoff, & D. Murdoch, *The Philosophical Writings of Descartes* (pp. 177–292). Cambridge, UK: Cambridge University Press.

de Sitter, W. (1917). On the Relativity of Inertia. Remarks Concerning Einstein's Latest Hypothesis. *Koninklijke Nederlandsche Akademie van Wetenschappen Proceedings, 19*(2), 1217–1225.

DeWitt, B. (1967). Quantum Theory of Gravity. I. Canonical Theory. *Physical Review, 160*(5), 1113–1148.

Dirac, P. A. (1927). The Quantum Theory of the Emission and Absorption of Radiation. *Proceedings of the Royal Society of London A, 114*(767), 243–265.

Dirac, P. A. (1928). The Quantum Theory of the Electron. *Proceedings of the Royal Society of London A, 117*(778), 610–624.

Dirac, P. A. (1930). *The Principles of Quantum Mechanics*. Oxford, UK: Oxford University Press.

Dirac, P. A. (1931). Quantised Singularities in the Electromagnetic Field. *Proceedings of the Royal Society of London A, 133*(821), 60–72.

Dirac, P. A. (1958). The Theory of Gravitation in Hamiltonian Form. *Proceedings of the Royal Society of London A, 246*(1246), 333–343.

Dirac, P. A. (1975). *General Theory of Relativity*. New York, NY: John Wiley & Sons.

DiSalle, R. (1994). On Dynamics, Indiscernibility, and Spacetime Ontology. *The British Journal for Philosophy of Science*, 265–287.

DiSalle, R. (2002). Newton's Philosophical Analysis of Space and Time. In I. B. Cohen & G. E. Smith, *The Cambridge Companion to Newton* (pp. 33–56). Cambridge, UK: Cambridge University Press.

DiSalle, R. (2006). *Understanding Space-Time.* Cambridge, UK: Cambridge University Press.

Dobbs, B. J. (1991). *The Janus Faces of Genius: The Role of Alchemy in Newton's Thought.* Cambridge, UK: Cambridge University Press.

Domski, M. (2010). Newton's Empiricism and Metaphysics. *Philosophy Compass, 5*(7), 525–534.

Douglas, V. A. (1956). *The Life of Arthur Stanley Eddington.* London, UK: Thomas Nelson.

du Châtelet, E. (1988 [1742]). *Institutions de Physique: Nouvelle edition.* (J. Ecole, Ed.). Hildesheim, Germany: Olms.

Duncan, A. (2012). *The Conceptual Framework of Quantum Field Theory.* Oxford, UK: Oxford University Press.

Duncan, A., & Janssen, M. (2008). Pascual Jordan's Resolution of the Conundrum of the Wave-Particle Duality of Light. *Studies in History and Philosophy of Modern Physics, 39*(3), 634–666.

Duncan, A., & Janssen, M. (2009). From Canonical Transformations to Transformation Theory, 1926–1927: The Road to Jordan's Neue Begründung. *Studies in History and Philosophy of Modern Physics, 40*(4), 352–362.

Duncan, A., & Janssen, M. (2012). (Never) Mind Your P's and Q's: Von Neumann versus Jordan on the Foundations of Quantum Theory. *European Journal of Physics H, 38*(2), 175–259.

Dürr, D., & Teufel, S. (2009). *Bohmian Mechanics: The Physics and Mathematics of Quantum Theory.* Berlin, Germany: Springer-Verlag.

Durrer, R. (2015). The Cosmic Microwave Background: The History of Its Experimental Investigation and Its Significance for Cosmology. *Classical and Quantum Gravity, 32*(12), 124007.

Earman, J. (1989). *World Enough and Space-Time: Absolute vs. Relational Theories of Space and Time.* Cambridge, MA: MIT Press.

Earman, J. (1995). *Bangs, Crunches, Whimpers, and Shrieks.* Oxford, UK: Oxford University Press.

Earman, J. (1999). The Penrose-Hawking Singularity Theorems: History and Implications. In H. Goenner, J. Renn, J. Ritter, & T. Sauer, *The Expanding Worlds of General Relativity* (pp. 235–267). Boston, MA: Birkhäuser.

Earman, J., & Eisenstaedt, J. (1999). Einstein and Singularities. *Studies in History and Philosophy of Modern Physics, 30*(2), 185–235.

Earman, J., & Glymour, C. (1980). Relativity and Eclipses: The British Eclipse Expeditions of 1919 and Their Predecessors. *Historical Studies in the Physical Sciences, 11*(1), 49–85.

Eddington, A. S. (1920). *Space, Time and Gravitation: An Outline of the General Relativity Theory.* Cambridge, UK: Cambridge University Press.

Einstein, A. (1905a). Über einen die Erzeugung und Verwandlung des Lichtes betreffenden heuristischen Gesichtspunkt. *Annalen der Physik, 17*(6), 132–148.

Einstein, A. (1905b). Über die von der molekularkinetischen Theorie der Wärme geforderte Bewegung von in ruhenden Flüssigkeiten suspendierten Teilchen. *Annalen der Physik, 17*(8), 549–560.

Einstein, A. (1905c). Zur Elektrodynamik bewegter Körper. *Annalen der Physik, 17*(10), 891–921.

Einstein, A. (1905d). Ist die Trägheit eines Körpers von seinem Energieinhalt abhängig? *Annalen der Physik, 18*(13), 639–641.

Einstein, A. (1906). Eine neue Bestimmung der Moleküldimensionen. *Annalen der Physik, 19*(4), 289–306.

Einstein, A. (1922). Ether and Relativity. In A. Einstein, *Sidelights on Relativity* (G. B. Jeffrey, & W. Perrett, Trans., pp. 3–24). London, UK: Meuthen & Co.

Einstein, A. (1924). Über den Äther. *Verhandlungen der Schweizerischen Naturforschenden Gesellschaft, 105*, 85–94.

Einstein, A. (1949). Autobiographical Notes. In P. A. Schilpp, *Albert Einstein: Philosopher-Scientist* (pp. 1–96). La Salle, IL: Open Court Press.

Einstein, A. (1998). *The Collected Papers of Albert Einstein* (Vol. 8). (R. Schulmann, A. J. Kox, M. Janssen, & J. Illy, Eds.). Princeton, NJ: Princeton University Press.

Einstein, A. (2005). *Einstein's Miraculous Year: Five Papers That Changed the Face of Physics.* (J. Stachel, Ed.). Princeton, NJ: Princeton University Press.

Einstein, A., Podolsky, B., & Rosen, N. (1935). Can Quantum-Mechanical Description of Physical Reality Be Considered Complete? *Physical Review, 47*(10), 777–780.

Einstein, A., & Rosen, N. (1937). On Gravitational Waves. *Journal of the Franklin Institute, 223*, 43–54.

Euler, L. (1750). Reflexions sur l'espace et le temps. *Memoires de l'academie des sciences de Berlin, 4*, 324–333.

Everett, H., III. (1957). "Relative State" Formulation of Quantum Mechanics. *Reviews of Modern Physics, 29*(3), 454–462.

Everitt, C. W. (1975). *James Clerk Maxwell: Physicist and Natural Philosopher.* New York, NY: Charles Scribner's Sons.

Farmelo, G. (2009). *The Strangest Man: The Hidden Life of Paul Dirac, Mystic of the Atom.* New York, NY: Basic Books.

Faye, J. (1991). *Niels Bohr: His Heritage and Legacy: An Anti-Realist View of Quantum Mechanics.* Dordrecht, Netherlands: Kluwer.

Faye, J., & Folse, H. J. (1994). *Niels Bohr and Contemporary Philosophy.* Dordrecht, Netherlands: Kluwer.

Feingold, M. (2004). *The Newtonian Moment: Isaac Newton and the Making of Modern Culture.* Oxford, UK: Oxford University Press.

Feintzeig, B. (2014). Hidden Variables and Incompatible Observables in Quantum Mechanics. *The British Journal for Philosophy of Science.* dx.doi.org/10.1093/bjps/axu017

Feynman, R. (1985a). *QED: The Strange Theory of Light and Matter.* Princeton, NJ: Princeton University Press.

Feynman, R. (1985b). *Surely You're Joking, Mr. Feynman.* New York, NY: W. W. Norton & Co.

Feynman, R. (1988). *"What Do You Care What Other People Think?": Further Adventures of a Curious Character.* New York, NY: W. W. Norton & Co.

Figala, K. (2002). Newton's Alchemy. In I. B. Cohen & G. E. Smith, *The Cambridge Companion to Newton* (pp. 370–386). Cambridge, UK: Cambridge University Press.

Fine, A. (1982a). Joint Distributions, Quantum Correlations, and Commuting Observables. *Journal of Mathematical Physics, 23*(7), 1306–1310.

Fine, A. (1982b). Hidden Variables, Joint Probability, and the Bell Inequalities. *Physical Review Letters, 48*(5), 291–295.

Fine, A. (1986). *The Shaky Game: Einstein, Realism, and the Quantum Theory.* Chicago, IL: University of Chicago Press.

FitzGerald, G. (1889). The Ether and Earth's Atmosphere. *Science, 13*, 390.

Fletcher, S. C. (2013). Light Clocks and the Clock Hypothesis. *Foundations of Physics, 43*(11), 1369–1383.

Fock, V. (1926). Zur Schrödingerschen Wellenmechanik. *Zeitschrift für Physik, 38*(3), 242–250.

Folse, H. J. (1985). *The Philosophy of Niels Bohr: The Framework of Complementarity.* Amsterdam, Netherlands: North-Holland.

Forbes, N., & Mahon, B. (2014). *Faraday, Maxwell, and the Electromagnetic Field.* Amherst, NY: Prometheus Books.

Frank, P. (1947). *Einstein: His Life and Times.* New York, NY: Alfred A. Knopf.

Fraser, D. (2008). The Fate of "Particles" in Quantum Field Theories with Interactions. *Studies in History and Philosophy of Modern Physics, 39*(4), 841–859.

Fraser, D. (2011). How to Take Particle Physics Seriously: A Further Defence of Axiomatic Quantum Field Theory. *Studies in History and Philosophy of Modern Physics, 42*(2), 126–135.

Fredenhagen, K., Rehren, K.-H., & Seiler, E. (2007). Quantum Field Theory: Where We Are. In I.-O. Stamatescu & E. Seiler, *Approaches to Fundamental Physics* (pp. 61–87). Berlin, Germany: Springer-Verlag.

Friedrichs, K. (1927). Eine Invariante Formulierun des Newtonschen Gravitationsgesetzes und der Grenzüberganges vom Einsteinschen zum Newtonschen Gesetz. *Mathematische Annalen, 98*, 566–575.

Gair, J. R., Vallisneri, M., Larson, S. L., & Baker, J. G. (2013). Testing General Relativity with Low-Frequency, Space-Based Gravitational-Wave Detectors. *Living Reviews in Relativity, 16*, 7.

Galison, P. (1979). Minkowski's Space-Time: From Visual Thinking to the Absolute World. *Historical Studies in the Physical Sciences, 10*, 85–121.

Galison, P. (1995). Theory Bound and Unbound: Superstrings and Experiment. In

F. Weinert, *Laws of Nature: Essays on the Philosophical, Scientific, and Historical Dimensions* (pp. 369–407). Berlin, Germany: De Gruyter.

Galison, P. (2000). The Suppressed Drawing: Paul Dirac's Hidden Geometry. *Representations, 72*, 145–166.

Galison, P. (2003). *Einstein's Clocks, Poincaré's Maps.* New York, NY: W. W. Norton & Co.

Galison, P., Gordin, M., & Kaiser, D. (2001). *Science and Society: Making Special Relativity.* New York, NY: Routledge.

Gambini, R., & Pullin, J. (2011). *A First Course in Loop Quantum Gravity.* Oxford, UK: Oxford University Press.

Garber, D. (1992). *Descartes' Metaphysical Physics.* Chicago, IL: University of Chicago Press.

Garber, D. (1995). Leibniz: Physics and Philosophy. In N. Jolley, *The Cambridge Companion to Leibniz* (pp. 270–352). Cambridge, UK: Cambridge University Press.

Garber, D. (2009). *Leibniz: Body, Substance, Monad.* Oxford, UK: Oxford University Press.

Garber, D. (2012). Leibniz, Newton, and Force. In A. Janiak & E. Schliesser, *Interpreting Newton: Critical Essays* (pp. 33–47). Cambridge, UK: University of Cambridge Press.

Georgi, H. (1993). Effective Field Theory. *Annual Review of Nuclear and Particle Science, 43*, 209–252.

Geroch, R. (1978). *General Relativity from A to B.* Chicago, IL: University of Chicago Press.

Ghirardi, G. C., Rimini, A., & Weber, T. (1986). Unified Dynamics for Microscopic and Macroscopic Systems. *Physical Review D, 34*(2), 470–491.

Gleick, J. (1992). *Genius: The Life and Science of Richard Feynman.* New York, NY: Pantheon.

Glick, T. F. (1987). *The Comparative Reception of Relativity.* Dordrecht, Netherlands: D. Reidel.

Glimm, J., & Jaffe, A. (1987). *Quantum Physics: A Functional Integral Point of View* (2nd ed.). Berlin, Germany: Springer-Verlag.

Goldman, M. (1983). *The Demon in the Aether: The Story of James Clerk Maxwell.* Edinburgh, UK: P. Harris.

Gordon, W. (1926). Der Comptoneffekt nach der Schrödingerschen Theorie. *Zeitschrift für Physik, 40*(1), 117–133.

Grant, E. (1981). *Much Ado about Nothing.* Cambridge, UK: Cambridge University Press.

Green, M. B., Schwarz, J. H., & Witten, E. (1987). *Superstring Theory.* Cambridge, UK: Cambridge University Press.

Greene, B. (1999). *The Elegant Universe: Superstrings, Hidden Dimensions, and the Quest for the Ultimate Theory.* New York, NY: W. W. Norton & Co.

Gross, D. J., & Wilczek, F. (1973). Ultraviolet Behavior of Non-Abelian Gauge Theories. *Physical Review Letters, 30*(26), 1343–1346.

Gubser, S. S. (2010). *The Little Book of String Theory.* Princeton, NJ: Princeton University Press.

Haag, R. (1992). *Local Quantum Physics.* Berlin, Germany: Springer-Verlag.

Haag, R. (1996). *Local Quantum Physics: Fields, Particles, Algebras* (2nd rev. ed.). Berlin, Germany: Springer-Verlag.

Hájek, A. (2012). *Interpretations of Probability.* Retrieved from the Stanford Encyclopedia of Philosophy: http://plato.stanford.edu/archives/win2012/entries/probability-interpret/

Hall, A. R. (1992). *Isaac Newton: Adventurer in Thought.* Oxford, UK: Blackwell.

Hall, A. R. (2002a). *Philosophers at War: The Quarrel between Newton and Leibniz.* Cambridge, UK: Cambridge University Press.

Hall, A. R. (2002b). Newton versus Leibniz: From Geometry to Metaphysics. In I. B. Cohen & G. E. Smith, *The Cambridge Companion to Newton* (pp. 431–454). Cambridge, UK: Cambridge University Press.

Halvorson, H., & Clifton, R. (2002). No Place for Particles in Relativistic Quantum Theories? In M. Kuhlmann, H. Lyre, & A. Wayne, *Ontological Aspects of Quantum Field Theory* (pp. 181–213). Singapore: World Scientific.

Hanson, N. R. (1961). Discovering the Positron. *The British Journal for the Philosophy of Science, XII*(47), 194–214.

Harman, P. M. (1998). *The Natural Philosophy of James Clerk Maxwell.* Cambridge, UK: Cambridge University Press.

Harper, W. (2012). *Isaac Newton's Scientific Method.* Oxford, UK: Oxford University Press.

Hartman, P. (1994). *A Memoir on the* Physical Review*: A History of the First Hundred Years.* New York, NY: American Institute of Physics.

Hawking, S. W., & Ellis, G. F. (1973). *The Large-Scale Structure of Space-Time.* Cambridge, UK: Cambridge University Press.

Hegerfeldt, G. C. (1974). Remark on Causality and Particle Localization. *Physical Review D, 10*(10), 3320–3321.

Hehl, F. W., & Obukov, Y. N. (2003). *Foundations of Classical Electrodynamics.* Boston, MA: Birkhäuser.

Heilbron, J. L. (1979). *Electricity in the 17th and 18th Centuries: A Study of Early Modern Physics.* Berkeley: University of California Press.

Heisenberg, W. (1925). Über quantentheoretische Umdeutung kinematischer und mechanischer Beziehungen. *Zeitschrift für Physik, 33*(1), 879–893.

Heisenberg, W. (1971). *Physics and Beyond: Encounters and Conversations.* New York, NY: Harper & Row.

Hendry, J. (1986). *James Clerk Maxwell and the Theory of the Electromagnetic Field.* Boston, MA: A. Hilger.

Hesse, M. (1961). *Forces and Fields: The Concept of Action at a Distince in the History of Physics.* London, UK: T. Nelson.

Hoddeson, L., Brown, L., Riordan, M., & Dresden, M. (1997). *The Rise of the Standard Model: A History of Particle Physics from 1964 to 1979*. Cambridge, UK: Cambridge University Press.

Hoefer, C. (1994). Einstein's Struggle for a Machian Gravitation Theory. *Studies in History and Philosophy of Science, Part A, 25*(3), 287–335.

Hoffman, B. (1972). *Albert Einstein: Creator and Rebel.* New York, NY: Viking Press.

Hollands, S. (2009). Axiomatic Quantum Field Theory in Terms of Operator Product Expansions: General Framework and Perturbation Theory via Hochschild Cohomology. *Symmetry, Integrability and Geometry: Methods and Applications, 5*, 90–134.

Holt, J. (2012). *Why Does the World Exist?* New York, NY: Liverlight Publishing Corporation.

Holton, G. (1988). *Thematic Origins of Scientific Thought: Kepler to Einstein* (rev. ed.). Cambridge, MA: Harvard University Press.

Hooker, C. A. (1972). The Nature of Quantum Mechanical Reality: Einstein versus Bohr. In R. G. Colodny, *Paradigms and Paradoxes: The Philosophical Challenge of the Quantum Domain* (pp. 67–302). Pittsburgh, PA: University of Pittsburgh Press.

Howard, D. (1985). Einstein on Locality and Separability. *Studies in History and Philosophy of Science Part A, 16*(3), 171–201.

Howard, D. (1990). Nicht sein kann was nicht sein darf: Or the Prehistory of EPR, 1909–1935. In A. I. Miller, *Sixty-Two Years of Uncertainty: Historical, Philosophical, Physics Inquiries into the Foundations of Quantum Physics* (pp. 61–111). New York, NY: Plenum.

Howard, D. (2014). Einstein and the Development of Twentieth-Century Philosophy of Science. In M. Janssen & C. Lehner, *The Cambridge Companion to Einstein* (pp. 354–376). Cambridge, UK: Cambridge University Press.

Huggett, N. (1994). What Are Quanta, and Why Does It Matter? In D. Hull, M. Forbes, & R. M. Burian, *PSA 1994: Proceedings of the Biennial Meeting of the Philosophy of Science Association* (pp. 69–76). Chicago, IL: University of Chicago Press.

Huggett, N. (2012). What Did Newton Mean by "Absolute Motion"? In A. Janiak & E. Schliesser, *Interpreting Newton: Critical Essays* (pp. 196–218). Cambridge, UK: Cambridge University Press.

Huggett, N., & Hoefer, C. (2015). *Absolute and Relational Theories of Space and Motion.* (E. N. Zalta, Ed.). Retrieved from the Stanford Encyclopedia of Philosophy (Spring 2015 Edition): http://plato.stanford.edu/archives/spr2015/entries/spacetime-theories/

Hughes, R. I. (1989). *The Structure and Interpretation of Quantum Mechanics.* Cambridge, MA: Harvard University Press.

Iliffe, R. (2007). *Newton: A Very Short Introduction.* Oxford, UK: Oxford University Press.

Isaacson, W. (2007). *Einstein: His Life and Universe.* New York, NY: Simon & Schuster.

Isham, C. J. (1992). *Canonical Quantum Gravity and the Problem of Time.* Retrieved from arXiv:gr-qc/9210011

Jackson, J. D. (1999). *Classical Electrodynamics* (3rd ed.). Hoboken, NJ: John Wiley & Sons.

Jammer, M. (1966). *The Conceptual Development of Quantum Mechanics.* New York, NY: McGraw Hill.

Jammer, M. (1974). *The Philosophy of Quantum Mechanics.* New York, NY: John Wiley & Sons.

Jammer, M. (1993). *Concepts of Space: The History of Theories of Space in Physics* (3rd ed.). Mineola, NY: Dover Publications.

Janiak, A. (2008). *Newton as Philosopher.* Cambridge, UK: Cambridge University Press.

Janssen, M. (2002a). COI Stories: Explanation and Evidence in the History of Science. *Perspectives on Science, 10*, 457–522.

Janssen, M. (2002b). Reconsidering a Scientific Revolution: The Case of Einstein versus Lorentz. *Physics in Perspective, 4*, 421–446.

Janssen, M. (2014). "No Success Like Failure . . .": Einstein's Quest for General Relativity, 1907–1920. In M. Janssen & C. Lehner, *The Cambridge Companion to Einstein* (pp. 167–227). Cambridge, UK: Cambridge University Press.

Janssen, M., & Renn, J. (2007). Untying the Knot: How Einstein Found His Way Back to Field Equations Discarded in the Zurich Notebook. In M. Janssen, J. D. Norton, J. Renn, T. Sauer, & J. Stachel, *The Genesis of General Relativity* (Vol. 2, pp. 839–925). Boston, MA: Birkäuser.

Janssen, M., & Renn, J. (2015). History: Einstein Was No Lone Genius. *Nature, 527*(7578), 298–300.

Jauch, J. M. (1968). *Foundations of Quantum Mechanics.* London, UK: Addison-Wesley Publishing Co.

Jauch, J. M., & Rohrlich, F. (1955). *The Theory of Photons and Electrons.* Reading, MA: Addison-Wesley Publishing Co.

Johnson, G. (2000, July). The Jaguar and the Fox. *The Atlantic*, 82–85.

Johnson, L. W., & Wolbarsht, M. L. (1979). Mercury Poisoning: A Probable Cause of Isaac Newton's Physical and Mental Ills. *Notes and Record of the Royal Society of London, 34*(1), 1–9.

Jolley, N. (2005). *Leibniz.* New York, NY: Routledge.

Jordan, P. (1927). Zur Quantenmechanik der Gasentartung. *Zeitschrift für Physik, 44*(6), 473–480.

Kaiser, D. (2005). *Drawing Theories Apart: The Dispersion of Feynman Diagrams in Postwar Physics.* Chicago, IL: University of Chicago Press.

Kennefick, D. (2007). *Traveling at the Speed of Thought: Einstein and the Quest for Gravitational Waves.* Princeton, NJ: Princeton University Press.

Kennefick, D. (2012). Not Only Because of Theory: Dyson, Eddington, and the

Competing Myths of the 1919 Eclipse Expedition. In C. Lehner, J. Renn, & M. Schemmel, *Einstein and the Changing Worldviews of Physics* (pp. 201–232). Boston, MA: Birkhäuser.

Kennefick, D. (2014). Einstein, Gravitational Waves, and the Theoretician's Regress. In M. Janssen & C. Lehner, *The Cambridge Companion to Einstein* (pp. 270–280). Cambridge, UK: Cambridge University Press.

Kent, S. L. (2001). *The Ultimate History of Video Games.* New York, NY: Three Rivers Press.

Kiefer, C. (2012). *Quantum Gravity* (3rd ed.). Oxford, UK: Oxford University Press.

Klein, O. (1926). Quantentheorie und fünfdimensionale Relativitätstheorie. *Zeitschrift für Physik, 37*(12), 895–906.

Kline, M. (1972). *Mathematical Thought from Ancient to Modern Times.* Oxford, UK: Oxford University Press.

Knox, E. (2014). Newtonian Spacetime Structure in Light of the Equivalence Principle. *The British Journal for Philosophy of Science, 65*(4), 863–888.

Kragh, H. (1981). The Genesis of Dirac's Relativistic Theory of Electrons. *Archive for History of Exact Sciences, 24*(1), 31–67.

Kragh, H. (1990). *Dirac: A Scientific Biography.* Cambridge, UK: Cambridge University Press.

Kragh, H. (2012). *Niels Bohr and the Quantum Atom: The Bohr Model of Atomic Structure 1913–1925.* Oxford, UK: Oxford University Press.

Krauss, L. (2011). *Quantum Man.* New York, NY: W. W. Norton & Co.

Krauss, L. M. (2012). *A Universe from Nothing.* New York, NY: Free Press.

Kuhlmann, M. (2015). *The History of QFT.* (E. N. Zalta, Ed.). Retrieved from the Stanford Encyclopedia of Philosophy: http://plato.stanford.edu/archives/sum2015/entries/quantum-field-theory/qft-history.html

Kuhn, T. (1978). *Black-Body Theory and the Quantum Discontinuity, 1894–1912.* Chicago, IL: University of Chicago Press.

Kuhn, T. S. (1962). The Historical Structure of Scientific Discovery. *Science, 136*(3518), 760–764.

Kursunoglu, B. N., & Wigner, E. P. (1990). *Paul Adrien Maurice Dirac: Reminiscences about a Great Physicist.* Cambridge, UK: Cambridge University Press.

Landsman, N. P. (2006). When Champions Meet: Rethinking the Bohr-Einstein Debate. *Studies in History and Philosophy of Modern Physics, 37*(1), 212–242.

Lehner, C. (2014). Einstein's Realism and His Critique of Quantum Mechanics. In M. Janssen & C. Lehner, *The Cambridge Companion to Einstein* (pp. 306–353). Cambridge, UK: Cambridge University Press.

Leibniz, G. W. (1989). *Philosophical Essays.* (R. Ariew & D. Garber, Eds.). Indianapolis, IN: Hackett Publishing Company.

Leibniz, G. W., & Clarke, S. (2000 [1715–16]). *Correspondence.* (R. Ariew, Ed.). Indianapolis, IN: Hackett Publishing Company.

Leone, M., & Robotti, N. (2010). Frédéric Joliot, Irène Curie and the Early History of the Positron. *European Journal of Physics, 31*, 975–987.

Lerche, W., Lüst, D., & Schellekens, A. N. (1987). Chiral Four-Dimensional Heterotic Strings from Self-Dual Lattices. *Nuclear Physics B, 287*, 477–507.

l'Hôpital, G. d. (1696). *Analyse des Infiniment Petits pour l'Intelligence des Lignes Courbes.* Paris, France: L'Imprimerie Royale.

Look, B. C. (2014). *Gottfried Willhelm Leibniz.* (E. N. Zalta, Ed.). Retrieved from the Stanford Encyclopedia of Philosophy: http://plato.stanford.edu/archives/spr2014/entries/leibniz/

Lorentz, H. A. (1892). De relatieve beweging van de aarde en den aether. *Verslagen van de gewone vergaderingen der Afdeling voor de Wis-en Natuurkundige Wetenschappen*, 74–79.

Lorentz, H. A. (1895). *Versuch einer Theorie der electrischen und optischen Erscheinungen in bewegten Körpern.* Leiden, Netherlands: E. J. Brill.

Lupher, T. (2010). Not Particles, Not Quite Fields: An Ontology for Quantum Field Theory. *Humana.mente, 13*, 155–173.

Maddy, P. (2007). *Second Philosophy: A Naturalistic Method.* Oxford, UK: Oxford University Press.

Mahon, B. (2003). *The Man Who Changed Everything: The Life of James Clerk Maxwell.* Hoboken, NJ: John Wiley & Sons.

Malament, D. B. (1996). In Defense of Dogma: Why There Cannot Be a Relativistic Quantum Mechanics of (Localizable) Particles. In R. Clifton, *Perspectives on Quantum Reality: Non-Relativistic, Relativisitc, and Field-Theoretic* (pp. 1–10). Dordrecht, Netherlands: Kluwer.

Malament, D. B. (2009, November). *Lecture Notes on Geometry and Space-Time.* Retrieved from http://www.socsci.uci.edu/-dmalamen/courses/geometryspacetimedocs/GST.pdf

Malament, D. B. (2012). *Topics in the Foundations of General Relativity and Newtonian Gravitation Theory.* Chicago, IL: University of Chicago Press.

Manuel, F. E. (1968). *A Portrait of Isaac Newton.* Cambridge, MA: Belknap Press.

Maudlin, T. (1994). *Quantum Non-Locality and Relativity.* Oxford, UK: Blackwell Publishing.

Maxwell, J. C. (1865). A Dynamical Theory of the Electromagnetic Field. *Philosophical Transactions of the Royal Society of London, 155*, 459–512.

Maxwell, J. C. (1891). *A Treatise on Electricity and Magnetism* (3rd ed.). Oxford, UK: Clarendon Press.

McDonough, J. K. (2014). *Leibniz's Philosophy of Physics.* (E. N. Zalta, Ed.). Retrieved from Stanford Encyclopedia of Philosophy: http://plato.stanford.edu/archives/spr2014/entries/leibniz-physics/

Mehra, J. (1994). *The Beat of a Different Drum: The Life and Science of Richard Feynman.* Oxford, UK: Clarendon Press.

Mehra, J., & Milton, K. A. (2000). *Climbing the Mountain: The Scientific Biography of Julian Schwinger.* Oxford, UK: Oxford University Press.

Mehra, J., & Rechenberg, H. (1982–2001). *The Historical Development of Quantum Theory.* New York, NY: Springer-Verlag.

Meli, D. B. (1993). *Equivalence and Priority: Newton versus Leibniz.* Oxford, UK: Oxford University Press.

Meli, D. B. (2002). Newton and the Leibniz-Clarke Correspondence. In I. B. Cohen & G. E. Smith, *The Cambridge Companion to Newton* (pp. 455–464). Cambridge, UK: Cambridge University Press.

Mercer, C. (2004). *Leibniz's Metaphysics: Its Origins and Development.* Cambridge, UK: Cambridge University Press.

Mercer, C., & Sleigh, J. R. (1995). Metaphysics: The Early Period to the Discourse on Method. In N. Jolley, *The Cambridge Companion to Leibniz* (pp. 67–123). Cambridge, UK: Cambridge University Press.

Meyer, H. W. (1972). *A History of Electricity and Magnetism.* Norwalk, CT: Burndy Library.

Michelson, A. A., & Morley, E. W. (1887). On the Relative Motion of the Earth and the Luminiferous Ether. *American Journal of Science, 34*, 333–345.

Miller, A. (1994). *Early Quantum Electrodynamics: A Sourcebook.* Cambridge, UK: Cambridge University Press.

Milonni, P. W. (1994). *The Quantum Vacuum: An Introduction to Quantum Electrodynamics.* San Diego, CA: Academic Press.

Minkowski, H. (1909). Raum und Zeit. *Physikalische Zeitschrift, 10*, 104–111.

Minkowski, H. (2012). *Space and Time: Minkowski's Papers on Relativity.* (V. Petkov, Ed., & F. L. Petkov, Trans.). Montreal, Québec: Minkowski Institute Press.

Misner, C. W., Thorne, K. S., & Wheeler, J. A. (1973). *Gravitation.* New York, NY: W. H. Freemen.

Mott, N. (1986). *A Life in Science.* London, UK: Taylor and Francis.

Moyer, D. F. (1981a). Origins of Dirac's Electron, 1925–1928. *American Journal of Physics*, 944–949.

Moyer, D. F. (1981b). Evaluations of Dirac's Electron, 1928–1932. *American Journal of Physics, 49*(11), 1055–1062.

Moyer, D. F. (1981c). Vindications of Dirac's Electron, 1932–1934. *American Journal of Physics, 49*(12), 1120–1125.

Murdoch, D. (1989). *Niels Bohr's Philosophy of Physics.* Cambridge, UK: Cambridge University Press.

Neffe, J. (2009). *Einstein: A Biography.* (S. Frisch, Trans.). Baltimore, MD: Johns Hopkins University Press.

Nersessian, N. (1984). *Farady to Einstein: Constructing Meaning in Scientific Theories.* Dordrecht, Netherlands: Martinus Nijhoff Publishers.

Newton, I. (1962). On the Gravity and Equilibrium of Fluids. In A. R. Hall & M. B. Hall, *Unpublished Scientific Papers of Isaac Newton* (pp. 89–156). Cambridge, UK: Cambridge University Press.

Newton, I. (1981). *The Mathematical Papers of Isaac Newton* (Vol. VIII: 1697–1722). (D. T. Whiteside, Ed.). Cambridge, UK: Cambridge University Press.

Newton, I. (1999, 1687/1713/1726). *The Principia: Mathematical Principles of Natural Philosophy.* (I. B. Cohen & A. Whitman, Eds.). Berkeley: University of California Press.

Norton, J. D. (1984). How Einstein Found His Field Equations: 1912–1915. *Historical Studies in the Physical Sciences, 14*, 253–316.

Norton, J. D. (2004). Einstein's Investigations of Galilean Covariant Electrodynamics Prior to 1905. *Archive for History of Exact Sciences, 59*, 45–105.

Norton, J. D. (2007). *How Did Einstein Think?* Retrieved from John Norton's Goodies Page: http://www.pitt.edu/-jdnorton/Goodies/Einstein_think/index.html

Norton, J. D. (2014). Einstein's Special Theory of Relativity and the Problems in the Electrodynamics of Moving Bodies That Led Him to It. In M. Janssen & C. Lehner, *The Cambridge Companion to Einstein* (pp. 72–102). Cambridge, UK: Cambridge University Press.

Odom, B., Hanneke, D., D'Urso, B., & Gabrielse, G. (2007). New Measurement of the Electron Magnetic Moment Using a One-Electron Quantum Cyclotron. *Physical Review Letters, 97*(3), 030801.

Pais, A. (1982). *Subtle Is the Lord: The Science and the Life of Albert Einstein.* Oxford, UK: Oxford University Press.

Pais, A. (1993). *Niels Bohr's Times: In Physics, Philosophy, and Polity.* Oxford, UK: Clarendon Press.

Pais, A. (1998). Paul Dirac: Aspects of His Life and Work. In A. Pais, M. Jacob, D. I. Olive, & M. F. Atiyah, *Paul Dirac: The Man and His Work* (pp. 1–45). Cambridge, UK: Cambridge University Press.

Palais, R. (1981). *The Geometrization of Physics.* Hsinchu, Taiwan: National Tsing Hua University.

Pashby, T. (2012). Dirac's Prediction of the Positron: A Case Study for the Current Realism Debate. *Perspectives on Science, 20*(4), 440–475.

Pauli, W. (1949). Einstein's Contributions to Quantum Theory. In P. A. Schilpp, *Albert Einstein: Philosopher-Scientist* (pp. 147–160). La Salle, IL: Open Court Press.

Peskin, M. E., & Schroeder, D. V. (1995). *An Introduction to Quantum Field Theory.* Boulder, CO: Westview Press.

Pickering, A. (1984). *Constructing Quarks: A Sociological History of Particle Physics.* Chicago, IL: University of Chicago Press.

Planck, M. (1900a). Ueber irreversible Strahlungsvorgänge. *Annalen der Physik, 306*(1), 69–122.

Planck, M. (1900b). Entropie und Temperatur strahlender Wärme. *Annalen der Physik, 306*(4), 719–737.

Planck, M. (1901). Ueber das Gesetz der Energieverteilung im Normalspectrum. *Annalen der Physik, 309*(3), 553–563.

Poincaré, H. (1906). Sur la dynamique de l'électron. *Rendiconti del Circolo matematico di Palermo, 21*, 129–176.

Polchinski, J. (1984). Renormalization and Effective Lagrangians. *Nuclear Physics B, 231*, 269–295.

Polchinski, J. (1992). Effective Field Theory and the Fermi Surface. Retrieved from arXiv:hep-th/9210046

Polchinski, J. (1998). *String Theory*. Cambridge, UK: Cambridge University Press.

Politzer, H. D. (1973). Reliable Perturbative Results for Strong Interactions? *Physical Review Letters, 30*(26), 1346–1349.

Pooley, O. (2013). Substantivalist and Relationalist Approaches to Spacetime. In R. Batterman, *The Oxford Handbook of Philosophy of Physics* (pp. 522–586). Oxford, UK: Oxford University Press.

Purcell, E. M. (1985). *Electricity and Magnetism* (2nd ed.). Boston, MA: McGraw-Hill.

Raffner, J. A. (2012). Newton's De Gravitatione: A Review and Reassessment. *Archive for History of the Exact Sciences*, 241–264.

Redhead, M. (1983). Quantum Field Theory for Philosophers. In P. D. Asquith & T. Nickles, *PSA 1982: Proceedings of the 1982 Biennial Meeting of the Philosophy of Science Association* (Vol. 2, pp. 57–99). Chicago, IL: University of Chicago Press.

Redhead, M. (1987). *Incompleteness, Nonlocality, and Realism*. Oxford, UK: Oxford University Press.

Redhead, M. (1988). A Philosopher Looks at Quantum Field Theory. In H. R. Brown & R. Harré, *Philosophical Foundations of Quantum Field Theory* (pp. 9–24). Oxford, UK: Clarendon Press.

Redhead, M. (1994). The Vacuum in Relativistic Quantum Field Theory. In D. Hull, M. Forbes, & R. M. Burian, *PSA 1994: Proceedings of the 1994 Biennial Meeting of the Philosophy of Science Association* (pp. 77–87). Chicago, IL: University of Chicago Press.

Redhead, M. (1995). More Ado about Nothing. *Foundations of Physics, 25*(1), 123–137.

Rejzner, K. (2016). *Perturbative Algebraic Quantum Field Theory*. Heidelberg, Germany: Springer.

Renn, J. (2007). *The Genesis of General Relativity*. Berlin, Germany: Springer.

Renn, J. (2013). Schrödinger and the Genesis of Wave Mechanics. In W. L. Reiter & J. Yngvason, *Erwin Schrödinger—50 Years After* (pp. 9–36). Zürich, Switzerland: European Mathematical Society.

Renn, J., & Sauer, T. (2007). Pathways out of Classical Physics. In M. Janssen, J. D. Norton, J. Renn, T. Sauer, & J. Stachel, *The Genesis of General Relativity* (Vol. 1, pp. 113–312). Boston, MA: Birkäuser.

Rickles, D. (2014). *A Brief History of String Theory: From Dual Models to M-Theory*. Berlin, Germany: Springer-Verlag.

Riles, K. (2013). Gravitational Waves: Sources, Detectors and Searches. *Progress in Particle and Nuclear Physics, 68*, 1–54.

Romer, M., & Cohen, I. B. (1940). Roemer and the First Determination of the Velocity of Light (1676). *Isis, 31*(2), 327–379.

Roque, X. (1997). The Manufacture of the Positron. *Studies in History and Philosophy of Modern Physics, 28*(1), 73–129.

Rovelli, C. (2001). *Notes for a Brief History of Quantum Gravity*. Retrieved from arXiv:gr-qc/0006061

Rovelli, C. (2004). *Quantum Gravity*. Cambridge, UK: Cambridge University Press.

Rugh, S. E., Zinkernagel, H., & Cao, T. Y. (1999). The Casimir Effect and the

Interpretation of the Vacuum. *Studies in History and Philosophy of Modern Physics*, 111–139.

Rynasiewicz, R. (1995a). By Their Properties, Causes and Effects: Newton's Scholium on Time, Space, Place and Motion—I. The Text. *Studies in History and Philosophy of Science, 26*(1), 133–153.

Rynasiewicz, R. (1995b). By Their Properties, Causes and Effects: Newton's Scholium on Time, Space, Place and Motion—II. The Context. *Studies in History and Philosophy of Science, 26*(2), 295–321.

Sabra, A. I. (1981). *Theories of Light, from Descartes to Newton.* Cambridge, UK: Cambridge University Press.

Sauer, T. (2014). Einstein's Unified Field Theory Program. In M. Janssen & C. Lehner, *The Cambridge Companion to Einstein* (pp. 281–305). Cambridge, UK: Cambridge University Press.

Sauer, T. (2015). Marcel Grossman and His Contribution to the General Theory of Relativity. In R. T. Jantzen, K. Rosquist, & R. Ruffini, *Proceedings of the Thirteenth Marcel Grossman Meeting on General Relativity* (pp. 456–503). Singapore: World Scientific.

Saunders, S. (1991). The Negative-Energy Sea. In S. Saunders & H. R. Brown, *The Philosophy of Vacuum* (pp. 65–109). Oxford, UK: Clarendon Press.

Saunders, S. (2002). Is the Zero-Point Energy Real? In M. Kuhlmann, H. Lyre, & A. Wayne, *Ontological Aspects of Quantum Field Theory* (pp. 313–344). Singapore: World Scientific.

Saunders, S. (2013). Rethinking Newton's Principia. *Philosophy of Science, 80*(1), 22–48.

Schaffner, K. F. (1972). *Nineteenth-Century Aether Theories.* Oxford, UK: Pergamon Press.

Schellekens, A. N. (2006). *The Landscape "avant la lettre."* Retrieved from arXiv: physics/0604134

Schrödinger, E. (1926a). Quantisierung als Eigenwertproblem. *Annalen der Physik, 385*(13), 437–490.

Schrödinger, E. (1926b). Quantisierung als Eigenwertproblem, II. *Annalen der Physik, 384*(6), 489–527.

Schrödinger, E. (1926c). Über das Verhältnis der Heisenberg-Born-Jordanschen Quantenmechanik zu der meinem. *Annalen der Physik, 384*(8), 734–756.

Schulmann, R., Kox, A. J., Janssen, M., & Illy, J. (1998). Editorial Note: The Einstein-De Sitter-Weyl-Klein Debate. In A. Einstein, R. Schulmann, A. J. Kox, M. Janssen, & J. Illy (Eds.), *The Collected Papers of Albert Einstein* (Vol. 8, pp. 351–357). Princeton, NJ: Princeton University Press.

Schwartz, M. D. (2013). *Quantum Field Theory and the Standard Model.* Cambridge, UK: Cambridge University Press.

Schwarzschild, K. (1916). Über das Gravitationsfeld eines Massenpunktes nach der Einsteinschen Theorie. *Sitzungsberichte der Königlich Preussischen Akademie der Wissenschaften, 7*, 189–196.

Schweber, S. S. (1994). *QED and the Men Who Made It: Dyson, Feynman, Schwinger, and Tomonaga.* Princeton, NJ: Princeton University Press.

Schwinger, J. (1969). *Particles, Sources, and Fields* (Vol. 1). New York, NY: Gordon Breach, Science Publishers.

Sciama, D. W. (1991). The Physical Significance of the Vacuum State of a Quantum Field. In S. Saunders & H. R. Brown, *The Philosophy of Vacuum* (pp. 137–158). Oxford, UK: Oxford University Press.

Sen, A. (1982). Gravity as a Spin System. *Physics Letters B, 119*(1–3), 89–91.

Shapin, S., & Schaffer, S. (1985). *Leviathan and the Air-Pump: Hobbes, Boyle, and the Experimental Life.* Princeton, NJ: Princeton University Press.

Shapiro, A. (2002). Newton's Optics and Atomism. In I. B. Cohen & G. E. Smith, *The Cambridge Companion to Newton* (pp. 227–255). Cambridge, UK: Cambridge University Press.

Shapiro, A. E. (2013). Newton's Optics. In J. Z. Buchwald & R. Fox, *The Oxford Handbook of the History of Physics* (pp. 166–198). Oxford, UK: Oxford University Press.

Shimony, A. (1963). Role of the Observer in Quantum Theory. *American Journal of Physics, 31*, 755–773.

Silverberg, R. (1965). *Niels Bohr: The Man Who Mapped the Atom.* Philadelphia, PA: Macrae Smith Company.

Slowik, E. (2014). *Descartes' Physics.* (E. N. Zalta, Ed.). Retrieved from the Stanford Encyclopedia of Philosophy: http://plato.stanford.edu/archives/sum2014/entries/descartes-physics/

Smeenk, C. (2014). Einstein's Role in the Creation of Relativistic Cosmology. In M. Janssen & C. Lehner, *The Cambridge Companion to Einstein* (pp. 228–269). Cambridge, UK: Cambridge University Press.

Smeenk, C., & Schliesser, E. (2013). Newton's Principia. In J. Z. Buchwald & R. Fox, *The Oxford Handbook of The History of Physics* (pp. 109–165). Oxford, UK: Oxford University Press.

Smith, G. (2008, Fall). *Isaac Newton.* (E. N. Zalta, Ed.). Retrieved from the Stanford Encyclopedia of Philosophy: http://plato.stanford.edu/archives/fall2008/entries/newton/

Smith, G. E. (2002). The Methodology of the Principia. In I. B. Cohen & G. E. Smith, *The Cambridge Companion to Newton* (pp. 138–173). Cambridge, UK: Cambridge University Press.

Smolin, L. (2001). *Three Roads to Quantum Gravity.* New York, NY: Basic Books.

Smolin, L. (2006). *The Trouble with Physics: The Rise of String Theory, the Fall of a Science, and What Comes Next.* Boston, MA: Houghton Mifflin Company.

Sorabji, R. (1988). *Matter, Space, and Motion: Theories in Antiquity and Their Sequel.* Ithaca, NY: Cornell University Press.

Spargo, P. E., & Pounds, C. A. (1979). Newton's "Derangement of the Intellect": New Light on an Old Problem. *Notes and Records of the Royal Society of London, 34*(1), 11–32.

Stachel, J. (1986). Einstein and the Quantum: Fifty Years of Struggle. In R. G. Colodny, *From Quarks to Quasars: Philosophical Problems of Modern Physics* (pp. 349–385). Pittsburgh, PA: University of Pittsburgh Press.

Stachel, J. (1987). Einstein and Ether Drift Experiments. *Physics Today, 40*, 45–47.

Stachel, J. (1989). Einstein's Search for General Covariance, 1912–1915. In D. Howard & J. Stachel, *Einstein and the History of General Relativity* (pp. 63–100). Boston, MA: Birkhäuser.

Stachel, J. (1999). The Early History of Quantum Gravity (1916–1940). In B. R. Iyer & B. Bhawal, *Black Holes, Gravitational Radiation and the Universe: Essays in Honor of C. V. Vishveshwara* (pp. 525–534). Dordrecht, Netherlands: Kluwer.

Stachel, J. (2005). Introduction. In A. Einstein, *Einstein's Miraculous Year: Five Papers That Changed the Face of Physics* (pp. 3–28). Princeton, NJ: Princeton University Press.

Stanley, M. (2003). "An Expedition to Heal the Wounds of War": The 1919 Eclipse and Eddington as Quaker Adventurer. *Isis, 94*(1), 57–89.

Stanley, M. (2007). *Practical Mystic: Religion, Science, and A. S. Eddington*. Chicago, IL: University of Chicago Press.

Stanton, R. (2015). *A Brief History of Video Games*. New York, NY: Little, Brown.

Stein, H. (1967). Newtonian Space-Time. *The Texas Quarterly, 10*, 174–200.

Stein, H. (1970). On the Notion of Field in Newton, Maxwell, and Beyond. In R. H. Stuewer, *Historical and Philosophical Perspectives of Science* (pp. 264–287). Minneapolis, MN: University of Minnesota Press.

Stein, H. (1972). On the Conceptual Structure of Quantum Mechanics. In R. Colodny, *Paradigms and Paradoxes: The Philosophical Challenge of the Quantum Domain* (pp. 367–438). Pittsburgh, PA: University of Pittsburgh Press.

Stein, H. (1977). Some Philosophical Prehistory of General Relativity. In J. Earman, C. Glymour, & J. Stachel, *Foundations of Space-Time Theories* (pp. 3–49). Minneapolis, MN: University of Minnesota Press.

Stein, H. (1991). On Relativity Theory and Openness of the Future. *Philosophy of Science, 58*(2), 147–167.

Stein, H. (2002). Newton's Metaphysics. In I. B. Cohen & G. E. Smith, *The Cambridge Companion to Newton* (pp. 256–307). Cambridge, UK: Cambridge University Press.

Stein, H. (Unpublished). Further Considerations on Newton's Methods. Retrieved from http://philsci-archive.pitt.edu/10633/

Stein, H. (Unpublished). On Metaphysics and Method in Newton. Retrieved from http://philsci-archive.pitt.edu/10631/

Stein, H. (Unpublished). Physics and Philosophy Meet: The Strange Case of Poincaré. Retrieved from http://philsci-archive.pitt.edu/10634/

Steinle, F. (2013). Electromagnetism and Field Physics. In J. Z. Buchwald & R. Fox, *The Oxford Handbook of the History of Physics* (pp. 533–570). Oxford, UK: Oxford University Press.

Stone, A. D. (2013). *Einstein and the Quantum: The Quest of the Valiant Swabian*. Princeton, NJ: Princeton University Press.

Streater, R. F. (1988). Why Should Anyone Want to Axiomatize Quantum Field Theory? In H. R. Brown & R. Harré, *Philosophical Foundations of Quantum Field Theory* (pp. 137–148). Oxford, UK: Oxford University Press.

Streater, R. F., & Wightman, A. S. (2000). *PCT, Spin and Statistics, and All That* (rev. ed.). Princeton, NJ: Princeton University Press.

Stroud, B. (1984). *The Significance of Philosophical Skepticism*. Oxford, UK: Oxford University Press.

Summers, S. J. (2011). Yet More Ado about Nothing: The Remarkable Relativistic Vacuum State. In H. Halvorson, *Deep Beauty: Understanding the Quantum World through Mathematical Innovation* (pp. 317–342). Cambridge, UK: Cambridge University Press.

Summers, S. J. (2012). *A Perspective on Constructive Quantum Field*. Retrieved from arXiv:1203.3991 [math-ph]

Susskind, L. (2003). The Anthropic Landscape of String Theory. In B. Carr, *Universe or Multiverse?* (pp. 247–266). Cambridge, UK: Cambridge University Press.

Susskind, L., & Friedman, A. (2014). *Quantum Mechanics: The Theoretical Minimum*. New York, NY: Basic Books.

Teller, P. (1993). Vacuum Concepts, Potentia, and the Quantum Field Theoretic Vacuum Explained for All. *Midwest Studies in Philosophy, 18*(1), 332–342.

Teller, P. (1995). *An Interpretive Introduction to Quantum Field Theory*. Princeton, NJ: Princeton University Press.

ter Haar, D. t. (1967). *The Old Quantum Theory*. Oxford, UK: Pergamon Press.

Tolstoy, I. (1981). *James Clerk Maxwell*. Edinburgh, UK: Canongate.

Trouton, F. T., & Noble, H. R. (1904). The Mechanical Forces Acting on a Charged Electric Condenser Moving through Space. *Philosophical Transactions of the Royal Society A, 202*, 165–181.

Tryon, E. P. (1973). Is the Universe a Vacuum Fluctuation? *Nature, 246*, 396–397.

Turnbull, H. W. (1959). *The Correspondence of Isaac Newton*. Cambridge, UK: Cambridge University Press.

Valente, G. (2015). Restoring Particle Phenomenology. *Studies in History and Philosophy of Modern Physics, 51*, 97–103.

van Dongen, J. (2010). *Einstein's Unification*. Cambridge, UK: Cambridge University Press.

van Fraassen, B. (1991). *Quantum Mechanics: An Empiricist View*. Oxford, UK: Oxford University Press.

van Vleck, J. H. (1972). Travels with Dirac in the Rockies. In A. Salam & E. P. Wigner, *Aspects of Quantum Theory* (pp. 7–16). Cambridge, UK: Cambridge University Press.

Voltaire. (1992 [1738]). *Eléments de la philosophie de Newton*. (R. Walters & W. Barber, Trans.). Oxford, UK: The Voltaire Foundation.

von Neumann, J. (1932). *Mathematische Grundlagen der Quantenmechanik*. Berlin, Germany: Springer.

Wald, R. M. (1984). *General Relativity*. Chicago, IL: University of Chicago Press.

Wald, R. M. (1994). *Quantum Field Theory in Curved Spacetime and Black Hole Thermodynamics*. Chicago, IL: University of Chicago Press.

Wallace, D. (2011). Taking Particle Physics Seriously: A Critique of the Algebraic Approach to Quantum Field Theory. *Studies in History and Philosophy of Modern Physics, 42*(2), 116–125.

Wallace, D. (2012). *The Emergent Multiverse: Quantum Theory According to the Everett Interpretation*. Oxford, UK: Oxford University Press.

Walter, S. A. (1999). Minkowski, Mathematicians and the Mathematical Theory of Relativity. In H. Goenner, J. Renn, J. Ritter, & T. Sauer, *The Expanding Worlds* (pp. 45–86). Boston, MA: Birkhäuser.

Warwick, A. (2003). *Masters of Theory: Cambridge and the Rise of Mathematical Physics*. Chicago, IL: University of Chicago Press.

Weatherall, J. O. (2011). On the Status of the Geodesic Principle in Newtonian and Relativistic Physics. *Studies in History and Philosophy of Modern Physics, 42*(4), 276–281.

Weatherall, J. O. (2013). *The Physics of Wall Street: A Brief History of Predicting the Unpredictable*. New York, NY: Houghton Mifflin Harcourt.

Weatherall, J. O. (2015). Fiber Bundles, Yang-Mills Theory, and General Relativity. *Synthese*. http://dx.doi.org/10.1007/s11229-015-0849-3

Weatherall, J. O. (2016a). Maxwell-Huygens, Newton-Cartan, and Saunders-Knox Spacetimes. *Philosophy of Science, 83*(1), 82–92.

Weatherall, J. O. (2016b). Regarding the "Hole Argument." *The British Journal for Philosophy of Science*. http://dx.doi.org/10.1093/bjps/axw012

Weatherall, J. O. (2016c). Are Newtonian Gravitation and Geometrized Newtonion Gravitation Theoretically Equivalent? *Erkenntnis*. http://dx.doi.org/10.1007/s10670-015-9783-5

Webb, J. (2013). *Nothing: From Absolute Zero to Cosmic Oblivion—Amazing Insights into Nothingness*. London, UK: Profile Books.

Weinberg, S. (1987). *Elementary Particles and the Laws of Physics, The 1986 Dirac Memorial Lectures*. Cambridge, UK: Cambridge University Press.

Weinberg, S. (1995–2000). *The Quantum Theory of Fields*. Cambridge, UK: Cambridge University Press.

Westfall, R. S. (1980). *Never at Rest*. Cambridge, UK: Cambridge University Press.

Weyl, H. (1922). *Space-Time-Matter* (4th ed.). (H. L. Brose, Trans.). London, UK: Methuen & Co.

Wheaton, B. R. (1983). *The Tiger and the Shark*. Cambridge, UK: Cambridge University Press.

Wheeler, J. A., & Feynman, R. P. (1945). Interaction with the Absorber as the Mechanism of Radiation. *Reviews of Modern Physics, 17*(2–3), 157–181.

Wheeler, J. A., & Feynman, R. P. (1949). *Reviews of Modern Physics, 21*(3), 425–433.

Whitaker, A. (1996). *Einstein, Bohr, and the Quantum Dilemma.* Cambridge, UK: Cambridge University Press.

Whittaker, E. T. (1951). *A History of the Theories of Aether and Electricity.* New York, NY: T. Nelson.

Wightman, A. (1972). The Dirac Equation. In A. Salam & E. P. Wigner, *Aspects of Quantum Theory* (pp. 95–116). Cambridge, UK: Cambridge University Press.

Wightman, A. (1989). The General Theory of Quantized Fields in the 1950s. In L. M. Brown, M. Dresden, & L. Hoddeson, *Pions to Quarks: Particle Physics in the 1950s* (pp. 608–629). Cambridge, UK: Cambridge University Press.

Wilson, K. G., & Kogut, J. (1974). The Renormalization Group and the ϵ Expansion. *Physics Reports, 12*(2), 75–199.

Wise, M. N. (2003). Pascual Jordan: Quantum Mechanics, Psychology, National Socialism. In M. Renneberg & M. Walker, *Science, Technology, and National Socialism* (pp. 224–254). Cambridge, UK: Cambridge University Press.

Woit, P. (2006). *Not Even Wrong.* New York, NY: Basic Books.

Wüthrich, A. (2010). *The Genesis of Feynman Diagrams.* Dordrecht, Nethrelands: Springer.

Index